Lecture Notes in Physics

New Series m: Monographs

The Editorial Policy for Monographs

The series Lecture Notes in Physics reports new developments in physical research and teaching - quickly, informally, and at a high level. The type of material considered for publication in the New Series m includes monographs presenting original research or new angles in a classical field. The timeliness of a manuscript is more important than its form, which may be preliminary or tentative. Manuscripts should be reasonably self-contained. They will often present not only results of the author(s) but also related work by other people and will provide sufficient motivation, examples, and applications.
The manuscripts or a detailed description thereof should be submitted either to one of the series editors or to the managing editor. The proposal is then carefully refereed. A final decision concerning publication can often only be made on the basis of the complete manuscript, but otherwise the editors will try to make a preliminary decision as definite as they can on the basis of the available information.
Manuscripts should be no less than 100 and preferably no more than 400 pages in length. Final manuscripts should preferably be in English, or possibly in French or German. They should include a table of contents and an informative introduction accessible also to readers not particularly familiar with the topic treated. Authors are free to use the material in other publications. However, if extensive use is made elsewhere, the publisher should be informed. Authors receive jointly 50 complimentary copies of their book. They are entitled to purchase further copies of their book at a reduced rate. As a rule no reprints of individual contributions can be supplied. No royalty is paid on Lecture Notes in Physics volumes. Commitment to publish is made by letter of interest rather than by signing a formal contract. Springer-Verlag secures the copyright for each volume.

The Production Process

The books are hardbound, and quality paper appropriate to the needs of the author(s) is used. Publication time is about ten weeks. More than twenty years of experience guarantee authors the best possible service. To reach the goal of rapid publication at a low price the technique of photographic reproduction from a camera-ready manuscript was chosen. This process shifts the main responsibility for the technical quality considerably from the publisher to the author. We therefore urge all authors to observe very carefully our guidelines for the preparation of camera-ready manuscripts, which we will supply on request. This applies especially to the quality of figures and halftones submitted for publication. Figures should be submitted as originals or glossy prints, as very often Xerox copies are not suitable for reproduction. In addition, it might be useful to look at some of the volumes already published or, especially if some atypical text is planned, to write to the Physics Editorial Department of Springer-Verlag direct. This avoids mistakes and time-consuming correspondence during the production period.
As a special service, we offer free of charge LaTeX and TeX macro packages to format the text according to Springer-Verlag's quality requirements. We strongly recommend authors to make use of this offer, as the result will be a book of considerably improved technical quality. The typescript will be reduced in size (75% of the original). Therefore, for example, any writing within figures should not be smaller than 2.5 mm.
Manuscripts not meeting the technical standard of the series will have to be returned for improvement.
For further information please contact Springer-Verlag, Physics Editorial Department V, Tiergartenstrasse 17, D-69121 Heidelberg, FRG.

Samuel D. Bogan   Mark K. Hinders

Interface Effects in Elastic Wave Scattering

Springer-Verlag
Berlin Heidelberg New York
London Paris Tokyo
Hong Kong Barcelona
Budapest

Authors

Samuel D. Bogan [1,2]
Mark K. Hinders [2,3]

[1] University of New Haven, College of Engineering
300 Orange Avenue, West Haven, CT 06516, USA

[2] Massachusetts Technological Laboratory, Inc.
330 Pleasant Street, Belmont, MA 02178, USA

[3] College of William and Mary, Applied Science
P. O. Box 8795, Williamsburg, VA 23187, USA

ISBN 3-540-57657-6 Springer-Verlag Berlin Heidelberg New York
ISBN 0-387-57657-6 Springer-Verlag New York Berlin Heidelberg

CIP data applied for.

Printed in Germany

SPIN: 10080557 2158/3140-543210 - Printed on acid-free paper

Dedication

To the memory of our friend and mentor, Professor Asim Yildiz.

Contents

List of Figures

Chapter 1

INTRODUCTION AND HISTORY

1.1 Motivation

Interface layers in composite materials exist as a natural consequence of material processing or are intentionally introduced into the composite to improve the chemical and physical properties of the composite. These interface layers for fiber or particle composites can be small, as in the interface between tungsten wire reinforcement and steel, or quite large, as in vapor deposited boron fibers in epoxy. These interfaces and have a direct influence on the macroscopic structural properties of the composite. Static structural properties due to these reinforcements can be characterized through experimental testing, and while static characterization is adequate for non-primary structural composites, composites are now commonly used as primary structural components, and dynamic properties are essential to the proper design of composite structures. The pertinent quantity of interest in dynamic analysis is the dynamic stress concentration, and the interface layers must be included in order to properly model the dynamical interaction with the substructure of the composite.

1.2 Overview

The interaction of an elastic wave with a layered spherical or cylindrical inclusion, elastic wave scattering, has been the focus of many authors concerned with non-destructive testing and prediction of dynamical effects in composite materials [51],[55],[73]. The difficulties encountered in measuring the properties of the interface between the fibers or particles and the matrix, make the analysis of these types of problems useful in predicting the bulk behavior of the composite material. Many different types of analysis have been made of these interfaces: e.g., finite element [73], integral methods [80], and eigenfunction approaches [51]. The latter is the only purely analytic method, the others require numerical matrix inversion at some point in the analysis. It is this eigenfunction method, or the method of Mie scattering, that is the focus of this analysis.

The origins of this method of analysis can be attributed to a number of authors who used the same techniques to analyze the interaction of light with optical lenses. At the time, 19th century and pre-Maxwell, the theory of light propagation was an elastic one. The pioneering work in this field was by Gustov Mie [41], A. Clebsch, Lord Rayleigh, and H. Lamb [1]–[4]. In these papers, Clebsch developed the general method of analysis and the exact solution for elastic wave scattering from a rigid, immovable sphere. Rayleigh showed with dimensional analysis that the energy scattered from a small sphere varies inversely as the fourth power of the wavelength (the Rayleigh Law), and Lamb investigated the scattering of incompressible elastic waves from small spherical and cylindrical scatterers.

Many closely related problems of the scattering of acoustic waves from small rigid spheres [5], elastic spheres and circular cylinders in viscous fluids [6], and acoustic wave scattering from fluid [7] and elastic spheres and circular cylinders of general size [8] have also been studied. In several well–known papers, R. Truell and his coworkers considered the scattering of compressional [9] and shear [10] waves from small elastic spheres, and longitudinal elastic wave scattering from an acoustic sphere [11]. The scattering of elastic waves from spherical cavities and rigid immovable spheres has also been studied by many authors [12]–[17]. In contrast to these researchers who have all used the general method of analysis of Clebsch, much of the recent work has concentrated on the numerical solution of integral or differential scattering equations [18]–[27].

The scattering of elastic waves on planar interfaces and the propagation of elastic waves through layered media has also been extensively studied. The reflection of elastic waves from an interface between two media was solved exactly by G. G. Stokes [28], and the transmission of plane elastic waves through a stratified medium containing alternate parallel planes of solid and liquid layers has been treated by R. B. Lindsay [30]. A matrix formulation for a stratified medium consisting of parallel solid elastic layers was given by W. T. Thomson [31] in a form convenient for numerical computations. Several textbooks are available [32] – [34] with extensive discussions of the work in this area. The study of the layered planar interface will provide essential insight into the layered spherical and layered cylindrical problem since the method of solution in all three will be shown here to be very similar.

The scattering of elastic waves from layered elastic inclusions of cylindrical and spherical geometry has been of interest only recently. While the acoustical problems of sound scattering from spherical and cylindrical shells has been the focus of much work over the last thirty years, from R.R. Goodman [36] to F. Leon [70], the treatment of elastic wave scattering with three separate elastic materials, exterior-shell-interior (see figure 1.1), has only recently been of greater interest.

The scalar problem of SH-elastic wave scattering from a two-layer elastic sphere has been solved [35], and a procedure for computing the solution to the n-layer problem presented [54]. We should note that there are three distinct elastic waves: one compressional, which can be thought of as an acoustic wave moving through the solid, and two distortional waves that are much like the transverse magnetic and transverse electric fields in electro-magnetism. When a boundary is encountered, these waves can all be coupled to one another. The SH-elastic wave is one of the distortional waves that does not interact with the compressional-longitudinal elastic wave for spherical boundaries. The solution therefore gives valuable insight into the more difficult fully coupled compressional-shear elastic wave scattering problem, but still leaves us with the more difficult problem. The more complicated compressional-distortional coupled (L-SV) problem of elastic wave scattering has been approached numerically by a number of authors, especially S.K. Datta who has studied the elastic wave interaction of thin spherical elastic shells in the long wave length approximation.[55] The analytic solutions for elastic wave scattering from a cylindrical or spherical inclusion is the necessary first step to studying the effects of an interface on the inclusion. The lack of these solutions, until recently[44]-[46], has limited studies to numerical or highly restrictive analytic solutions.

In this monograph, we derive the exact solutions for elastic wave scattering from spherical, cylindrical, and planar elastic interfaces. The method of approach is the exact theory of Clebsch [1] and Mie [41] which has been recently used to derive the exact solution for plane elastic wave scattering from spherical and cylindrical elastic inclusions in an infinite elastic medium [44] - [46]. The layered-inclusion problem can be readily formulated, but difficulties arise since the resulting system of boundary condition equations is 8×8 and an analytic solution becomes complicated. In order to overcome this difficulty we will employ the *MATHEMATICA* symbolic manipulation software package to check the tedious algebraic manipulations that must by performed by hand. In addition, we will be comparing a numerical solution for specific cases to confirm the analytic solutions. The subset of solutions for elastic wave scattering by non-layered planar, cylindrical, and spherical interfaces will also be used as a confirmation of the solution.

Having obtained the exact solution for elastic wave scattering from layered planar, cylindrical, and spherical interfaces, we will then study the dynamic stress concentrations and their relation to the physical parameters. The motivation and focus of this work is to establish the foundation for predicting the measurable backscattered ultrasonic field during non-destructive evaluation of composites and to predict potentially damaging stress concentrations caused by dynamical interactions with the interface layer. These studies are easily performed by adjusting parameters in the general expressions. Since there are no restrictions placed on the ranges of the material parameters, the core can be considered a vacuum or a rigid material by setting that medium's density to small values or very large values respectively. Similarly, considering the infinite medium to be an inviscid fluid instead of a solid elastic medium by setting the shear modulus near zero, we can then study acoustic wave scattering from a two-layer sphere of various configurations [83]. Many other special cases follow in the same manner, all found directly from the general solution by allowing certain parameters to take on extreme values. Two important special cases are when the scatterer is small compared to the wavelength (Rayleigh scattering), and when the shell and core materials are nearly identical. In this latter case, we recover the known results for a non-layered sphere or cylinder.

Numerical computation of the general solution is accomplished with relative ease since we need only *evaluate* functions and sum series. No complicated calculations such as numerical matrix inversion or numerical integration are required since we will already have an analytic expression for the result. The scattering behavior can thus be readily characterized via variation of frequency, radii, and material parameters over wide ranges.

1.3 The pioneering work of Clebsch and Mie

In 1861, Alfred Clebsch wrote what could have been the classic reference for problems in dynamical elasticity[1]. Motivated by the hope of improving on the practice of designing mirrors by the laws of geometric optics, Clebsch used the theory of the elastic "aether" to formulate and solve what we would classify today as the scattering of an elastic wave by a perfectly rigid elastic sphere. In his analysis, Clebsch derived the appropriate series solutions for the problem, discovering what will later be called the Debye potentials, spherical harmonics, and the Rayleigh law ten years prior to Rayleigh's famous work on the subject [81]. Though the work of Clebsch was published in the prominent journal of the time, his work was largely overlooked by later authors. This was in part due to the introduction of Maxwell's equations and perhaps to it being his last work prior to his early death.

Gustov Mie is generally credited with the eigenfunction technique for solving wave scattering problems. His work on light scattering from colloidal gold solutions gained general acceptance, probably due to its application to the brilliant scattering effects of light in stained glass. Coupling the theoretical framework with a well known effect earned the attention that over the last century has tied his name to this technique[41]. Lord Rayleigh, using only dimensional analysis, showed that the scattering of light in the atmosphere was dominated by a simple power law, providing an explanation for the blue hue of the sky [81].

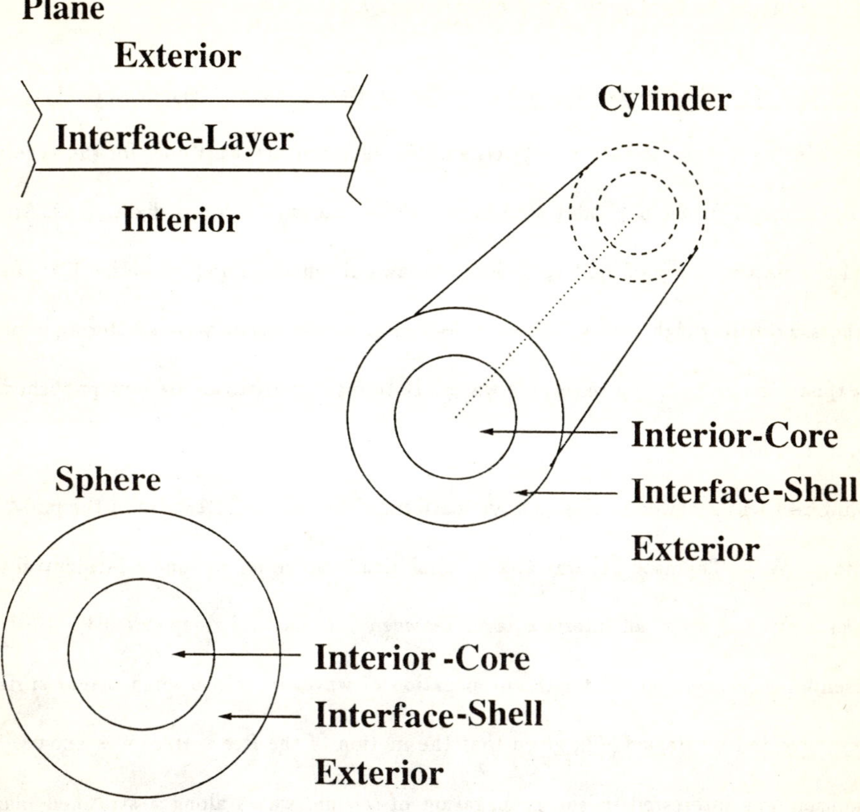

Figure 1.1: Region definitions for differing geometries.

1.4 Planar interfaces and geophysics

In the case of an elastic wave incident upon the interface between two media, the historical path leads to the mathematical physicists of the 19th century, as it does for the spherical and cylindrical problems. Interested in explaining the diffraction of light, G.G. Stokes used the elastic theory of light to derives the laws of reflection and refraction [28]. Later authors, Lord Rayleigh and A.E.H. Love, looked at planar elastic wave scattering in order to explain the propagation of seismic waves. Both historic treatises are now published by Dover [81],[82].

Interest was renewed in elastic wave scattering from planar interfaces in the mid-20th century. W.T. Thomson[31] was able to analytically solve for obliquely incident elastic waves scattering from an interface layer between two elastic half-spaces [64]. He later presented a general method for the propagation of waves in an n-layered planar surface, later corrected by Haskel [79], given that the motion of the free surface was known [31]. Thomson was interested in the propagation of seismic waves along a stratified planar elastic medium, and the majority of the literature deals with this type of free surface or specified surface condition problem.

The problem of interest in the non-destructive evaluation (NDE) of laminated composites is different since the incident field is known, but none of the displacements or stresses on the boundaries can be specified. This is the problem of present interest in the evaluation of laminated composites by NDE and it is this more general problem that is solved in this work.

1.5 Spheres, cylinders, and mid 20th century physics

In the late 19th century, an incredible upsurge of interest in optical diffraction problems drew in the finest minds in mathematics, from the initial work of Stokes [28] and Clebsch [1] to later work by Rayleigh [81], Mie [41], and L.V. Lorenz [2], the Danish physicist, not to be confused with H.A. Lorentz. This work is very similar in approach to the techniques used for solving the problems of elastic wave scattering from elastic spheres and cylinders. We therefore credit these brilliant mathematical physicists with founding the current field of elastic wave scattering and stimulating the natural engineering outgrowth in the mid-20th century.

A re-emergence of this technique to solving scattering problems occurred in the 1950's. One of the first papers was that of F.Fox [58] on the acoustic pressure on a sphere. Later works include the well respected work of Johnson & Truell [52], computing elastic scattering cross-sections for numerous elastic materials. During this time, the Mie scattering approach to solving acoustic and elastic scattering problems became a prominent feature in the "Journal of the Acoustical Society of America" and "Journal of Applied Physics". The incredible variety of problems that could be approached by this technique, from the analysis of the interaction of shearing waves with buried cylinders studied by C.C Mow and L.J Mente to the work of P. Epstein and R. Carhart analyzing acoustic interaction with fog, makes it of enormous use to applied physicists and engineers alike. The most prominent journal articles in sound wave scattering were those of V.C. Anderson for sound scattering from a fluid sphere [7], J.J. Faran [8] analyzing sound scattering from both elastic spheres and cylinders, extended by M.C Junger [39] to include the effects of an elastic shell in both geometries, and R.M. White who studied obliquely incident elastic waves on cylinders analytically and experimentally.

Since the early 60's, there have been many related papers and texts dealing with these types of problems. The focus has been on the computational or restrictive analytic solutions rather than developing general analytic solutions. It is not until we look further ahead to the literature that we find analytic expressions to these boundary value problems [44],[46]. The recent development of symbolic manipulators, computer programs for algebraic manipulation, make verification of these complex problems possible, although the independent solution by symbolic manipulators without direction and human insight is still beyond the capacities of present systems. Just as the numerical results of Johnson and Truell [52] have been rigorously checked, lengthy analytic solutions can now be verified with the use of symbolic manipulators. This makes possible the analysis of the modern complex problems of interfaces in composite materials.

1.6 Modern studies of composites

The introduction of a wide variety of composite materials in structural applications makes the dynamical analysis of composite materials of great interest to mechanical engineers. Whether for ultrasonic analysis or calculation of dynamic stress concentrations, the need for a dynamical analysis of these materials is clear. Some of the most closely related work of other authors is that of S.K. Datta [55], A.I. Beltzer [77], and M.K. Hinders [45]. In one of his papers, S.K. Datta analyzed the interaction between an ultrasonic wave and an ellipsoidal crack in order to establish a method for characterizing cracks by nondestructive analysis (NDE) [55]. In a later paper, Datta studied the attenuation and phase relations in particle reinforced composites for characterization and prediction of overall dynamical properties [56]. In his work with P. Olsson, an interface was added around the inclusions to study the effect on the scattering cross-sections [51]. A.I. Beltzer studied the elastic wave interactions with cylinders and spheres to predict stress concentrations in composites, though only accounting for the uncoupled shear wave, and M.K. Hinders et al. [45] have studied the analytic solutions to cylindrical and spherical wave scattering in order to predict dynamic stress concentrations.

There have been numerous papers on this topic. The major objectives are the general problem of predicting the dynamic behavior of the composite as it interacts at the reinforcement level with the compressional and distortional wave, or to predict the backscattered fields from a known configuration as a benchmark for the ultrasonic characterization of composite materials. Both areas of interests can be studied with the method of Mie.

In this analysis, we are studying the dynamical effects of incident compressional and distortional elastic waves on a layered plane, cylinder, and sphere. The focus is on the stress fields immediately surrounding the interface layer in the matrix. The motivation is the current extensive and growing use of reinforced composite materials in applications where dynamical effects may have a significant effect on the lifetime of the composite. As an example, boron and silicon-carbide fibers manufactured by chemical vapor deposition (CVD) are actually cylindrical shells of boron and silicon-carbide with core substrates of either tungsten or carbon. Thus the fiber-reinforced composites are actually cylindrical-shell fiber reinforcements in a metal or polymer matrix. Another example of the distinct shell structure is tungsten wire in steel. While there is not an intentional shell structure introduced as in CVD, an interface layer forms between the tungsten and steel that should be modeled in addition to the fiber and matrix to properly characterize the material.

Chapter 2

ELASTIC MODEL AND FORMULATION

2.1 Equations of dynamical elasticity

We next outline the formulation of the elastic scattering problem for compressional (longitudinal) and shearing (transverse) plane elastic waves from a planar interface layer separating two elastic half-spaces, a shelled circular cylinder surrounded by an exterior medium, and a shelled sphere in an exterior medium (see figure 2.1).

We assume that the elastic medium in which the layered object is embedded extends to infinity, and that the exterior, interface layer, and interior medium can all be described by the equations of motion with differing material parameters. In setting up the analytic model, we use the dynamical equation of motion for a homogeneous and isotropic medium as derived in appendix A:

$$\rho\partial_t^2\vec{u} - \mu\nabla^2\vec{u} - (\mu + \lambda)\nabla(\nabla \cdot \vec{u}) = 0 \tag{2.1}$$

where ρ is the medium density, μ and λ are elastic Lamé parameters, and $\vec{u}$ is the displacement vector. After some manipulation and assuming purely harmonic time dependence $e^{-i\omega t}$, the equation of motion can be written as

$$(\nabla^2 + K^2)\vec{u} - (1 - \frac{K^2}{k^2})\nabla(\nabla \cdot \vec{u}) = 0 \tag{2.2}$$

Here, $K = \omega/c_T$ and $k = \omega/c_L$ are the propagation constants for transverse and longitudinal elastic waves, with $c_T^2 = \mu/\rho$ and $c_L^2 = (\lambda + 2\mu)/\rho$ defining the transverse and longitudinal wave propagation velocities respectively. Longitudinal and transverse waves can be immediately separated by using a Helmholtz vector field decomposition with $\vec{u} = \vec{u}_L + \vec{u}_T$, where the longitudinal field satisfies $\nabla \times \vec{u}_L = 0$ and the transverse field satisfies $\nabla \cdot \vec{u}_T = 0$. The longitudinal displacement field is irrotational with displacements along the direction of propagation. The transverse displacement field is solenoidal with displacements perpendicular to the direction of propagation, distorting or shearing the medium as it propagates. The total field, via Helmholtz, is a linear combination of these fields. A further separation of the transverse field is possible in cartesian, cylindrical, and spherical coordinate systems, if we separate the transverse or shear field into two potential fields denoted by π_S and π_V and related to the transverse displacement field by:

$$\vec{u}_T = \frac{-i}{K}\nabla \times (\vec{a}\pi_S) - \left(\frac{i}{K}\right)^2 \nabla \times \nabla \times (\vec{a}\pi_V) \tag{2.3}$$

where $\vec{a}$ is a vector specific to the coordinate system in the problem. The longitudinal or compressional displacement field can similarly be written in terms of a longitudinal displacement potential π_L given by:

$$\vec{u}_L = \frac{-i}{k}\nabla \pi_L \tag{2.4}$$

The complex factors multiplying each of the displacement potentials are for normalization, equating the magnitude of the displacement field to unity if we were to choose π_L, π_S, or π_V as the displacement potential for a plane wave. The three potentials when substituted into the equation of motion result in three scalar Helmholtz equations.

$$(\nabla^2 + k^2)\pi_L = 0, \quad (\nabla^2 + K^2)\pi_S = 0, \quad (\nabla^2 + K^2)\pi_V = 0 \tag{2.5}$$

The solutions to these three scalar Helmholtz equations in cartesian, cylindrical, and spherical coordinates are well known and will be used to relate the displacement potentials to the displacement and stress fields. If we assign the subscripts 1,2,3 to the elastic Lamé parameters and the density of the outside medium, the interface layer, and the interior as shown in figure 2.1, we can identify the distinct regions or fields that must be described by the displacement potentials. For each region, we will have three potentials that will describe the displacements within that region. In interface regions each potential will be a linear combination of two potentials giving a total of six potentials for an interface region. The three problem geometries of interest in this book will require six displacement potentials to describe the interface, three displacement potentials for the exterior, and three to describe the interior. This is the most general situation for three distinct regions involving a total of twelve displacement potentials to solve the problem.

2.1.1 Cartesian coordinates

In the first case, a planar elastic layer separates two elastic half spaces, modeling an interface in a laminated composite. The solutions to the scalar Helmholtz equation, as discussed in appendix B, provide functions to substitute into the displacement potentials for calculating displacements and stresses. These solutions are given by the exponentials:

$$\pi_L = e^{i\vec{k}\cdot\vec{z}}e^{i\vec{k}\cdot\vec{x}}, \qquad \pi_S = e^{i\vec{K}\cdot\vec{z}}e^{i\vec{K}\cdot\vec{x}}, \qquad \pi_V = e^{i\vec{K}\cdot\vec{z}}e^{i\vec{K}\cdot\vec{x}} \tag{2.6}$$

where $\vec{k}$ and $\vec{K}$ are the wave numbers multiplied by a normalized vector pointing in the direction of propagation. The direction of propagation of each of the waves in each of the fields follows very simply from the laws of reflection and refraction otherwise known as Snell's law of refraction given by:

$$\vec{k_1}\cdot\hat{x} = k_1 \sin\phi_L^s = \vec{k_2}\cdot\hat{x} = k_2\sin\phi_L^r = \vec{k_3}\cdot\hat{x} = k_3\sin\phi_L^t$$

$$= \vec{K_1}\cdot\hat{x} = K_1\sin\phi_T^s = \vec{K_2}\cdot\hat{x} = K_2\sin\phi_T^r = \vec{K_3}\cdot\hat{x} = K_3\sin\phi_T^t \tag{2.7}$$

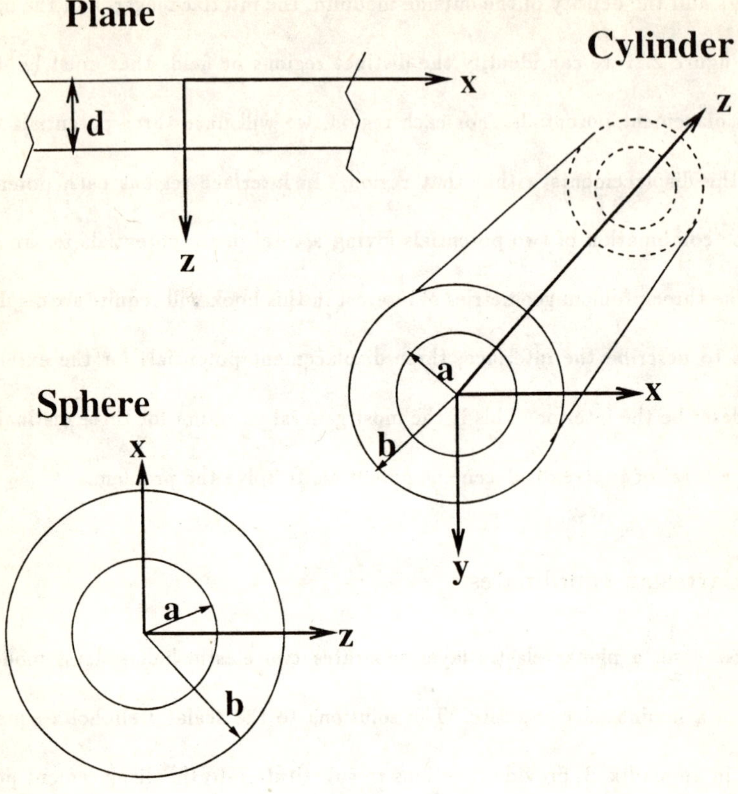

Figure 2.1: Problem geometry for scattering from elastic interface layers.

The angles ϕ are the acute angles between the propagation vector and the $\hat{z}$ axis, $K_i^2 = \omega^2\rho_i/\mu_i$ and $k_i^2 = \omega^2\rho_i/(\lambda_i + 2\mu_i)$ for $i = (1,2,3)$. The primary direction of propagation is taken as in the positive $\hat{z}$ and $\hat{x}$ direction. For most choices of elastic material parameters, there are angles that will invalidate the above relationship. In that case, it is understood that k_i and K_i are complex and the above relations are modified to the hyperbolic trigonometric functions.

2.1.2 Cylindrical coordinates

In cylindrical coordinates, the solution to the Helmholtz equations for the displacement potentials of the displacement fields are given by:

$$\begin{aligned}
\pi_L &= \sum_{n=0}^{\infty} \left[\begin{array}{c} J_n(\vec{k}\cdot\vec{r}) \\ H_n^{(1,2)}(\vec{k}\cdot\vec{r}) \end{array} \right] e^{in\theta} e^{i\vec{k}\cdot\vec{z}} \\
\pi_S &= \sum_{n=0}^{\infty} \left[\begin{array}{c} J_n(\vec{K}\cdot\vec{r}) \\ H_n^{(1,2)}(\vec{K}\cdot\vec{r}) \end{array} \right] e^{in\theta} e^{i\vec{K}\cdot\vec{z}} \qquad (2.8) \\
\pi_V &= \sum_{n=0}^{\infty} \left[\begin{array}{c} J_n(\vec{K}\cdot\vec{r}) \\ H_n^{(1,2)}(\vec{K}\cdot\vec{r}) \end{array} \right] e^{in\theta} e^{i\vec{K}\cdot\vec{z}}
\end{aligned}$$

where $\vec{k}$ and $\vec{K}$ are the wave numbers multiplied by a base vector pointing in the direction of propagation and Snell's law of refraction is used to determine the propagation direction in each of the three media. The displacement potentials are Fourier-Bessel series multiplying an exponential function containing the axial ($\hat{z}$) dependence, and the choice of the radial function is determined by the boundary conditions of the field represented by the potentials. A common addition to the potentials is ϵ_n, where $\epsilon_0 = 1$ and $\epsilon_n(n \neq 0) = 2$, and i^n. This anticipates the expansion of an incident planar displacement wave that will contains an ϵ_n and i^n, and the subsequent solution for the modal coefficients that will also contain them. Extracting them in advance from the coefficients of the displacement potentials simplifies the later algebra.

2.1.3 Spherical coordinates

In spherical coordinates, the solutions to the Helmholtz equation for the displacement potentials describing the displacement fields are the well known Ricatti functions and spherical harmonics. After some reductions for orthogonality of the series solutions the expressions for the displacement potentials of the sphere are:

$$
\begin{aligned}
(kr)\pi_L &= \sum_{l=0}^{\infty} \begin{bmatrix} \psi_l(kr) \\ \zeta_l^{(1,2)}(kr) \end{bmatrix} P_l(\cos\theta) \\
(Kr)\pi_S &= \sum_{l=1}^{\infty} \begin{bmatrix} \psi_l(Kr) \\ \zeta_l^{(1,2)}(Kr) \end{bmatrix} P_l^{(1)}(\cos\theta)\sin\phi \\
(Kr)\pi_V &= \sum_{l=1}^{\infty} \begin{bmatrix} \psi_l(Kr) \\ \zeta_l^{(1,2)}(Kr) \end{bmatrix} P_l^{(1)}(\cos\theta)\cos\phi
\end{aligned}
\tag{2.9}
$$

The choice of the radial function is determined by the boundary conditions of the field represented by the displacement potentials. The addition of kr in the potentials allows us to use the Ricatti-Bessel functions rather than the spherical Bessel functions. It should be remembered that the potentials that satisfy the scalar Helmholtz equation are the π's and not the $kr\pi$'s. Ricatti functions are defined as

$$
\begin{bmatrix} \psi_l(kr) \\ \zeta_l^{(1,2)}(kr) \end{bmatrix} = \begin{bmatrix} \sqrt{\pi kr/2}J_{l+1/2}(kr) \\ \sqrt{\pi kr/2}H_{l+1/2}^{(1,2)}(kr) \end{bmatrix} = \begin{bmatrix} (kr)j_l(kr) \\ (kr)h_l^{(1,2)}(kr) \end{bmatrix} \tag{2.10}
$$

where $J_{l+1/2}(kr)$ and $H_{l+1/2}^{(1,2)}(kr)$ are the half-order cylindrical Bessel and Hankel functions and $j_l(kr)$ and $h_l^{(1,2)}(kr)$ are the spherical Bessel and Hankel functions.

2.2 Boundary conditions

Conditions that hold at a surface separating two elastic media are easily derivable from the field equations in appendix A. At such a surface, the requirement that the two media remain in perfect contact leads to the conclusion that the displacements must be continuous across the boundary. The equilibrium of an arbitrarily small volume which encloses portions of both media leads to the continuity of normal surface tractions across the surface. These conditions are sometimes referred to as the welded contact boundary conditions. One advantage to the modeling of an interface layer is that given a thin interface layer, the interface can take on properties that resemble debonds or delaminations or odd boundary conditions that would otherwise have to be introduced into the field equations in a non-physical way.

The displacements, as previously defined in section 2.1 and appendix A, are written in terms of the displacement potentials as:

$$\vec{u} = \frac{-i}{k}\nabla\pi_L + \frac{-i}{K}\nabla\times(\vec{a}\pi_S) - \left(\frac{i}{K}\right)^2\nabla\times\nabla\times(\vec{a}\pi_V) \tag{2.11}$$

The three coordinate systems under study are orthogonal curvilinear, and the displacements and normal surface tractions can be written in a general orthogonal curvilinear form. If we define $\vec{a} = f_1\hat{e}_1$, where $f_1\hat{e}_1$ is $(\hat{e}_y, \hat{e}_z, r\hat{e}_r)$, then the displacements and normal surface tractions can be written for cartesian, cylindrical, and spherical coordinates as follows:

$$
\begin{aligned}
u^1 &= \frac{-i}{k\sqrt{g_{11}}}\frac{\partial}{\partial x^1}(f_1\pi_L) + 0 - \left(\frac{i}{K}\right)^2 \frac{-1}{\sqrt{g_{22}g_{33}}}\left[\frac{\partial}{\partial x^2}\left(\sqrt{\frac{g_{33}}{g_{11}g_{22}}}\frac{\partial}{\partial x^2}(\sqrt{g_{11}}f_1\pi_V)\right) + \frac{\partial}{\partial x^3}\left(\sqrt{\frac{g_{22}}{g_{11}g_{33}}}\frac{\partial}{\partial x^3}(\sqrt{g_{11}}f_1\pi_V)\right)\right] \\
u^2 &= \frac{-i}{k\sqrt{g_{22}}}\frac{\partial}{\partial x^2}(f_1\pi_L) + \frac{-i}{K\sqrt{g_{11}g_{33}}}\frac{\partial}{\partial x^3}(\sqrt{g_{11}}f_1\pi_S) - \left(\frac{i}{K}\right)^2 \frac{1}{\sqrt{g_{11}g_{33}}}\frac{\partial}{\partial x^1}\left(\sqrt{\frac{g_{33}}{g_{11}g_{22}}}\frac{\partial}{\partial x^2}(\sqrt{g_{11}}f_1\pi_V)\right) \\
u^3 &= \frac{-i}{k\sqrt{g_{33}}}\frac{\partial}{\partial x^3}(f_1\pi_L) + \frac{+i}{K\sqrt{g_{11}g_{22}}}\frac{\partial}{\partial x^2}(\sqrt{g_{11}}f_1\pi_S) - \left(\frac{i}{K}\right)^2 \frac{1}{\sqrt{g_{11}g_{22}}}\frac{\partial}{\partial x^1}\left(\sqrt{\frac{g_{22}}{g_{11}g_{33}}}\frac{\partial}{\partial x^3}(\sqrt{g_{11}}f_1\pi_V)\right) \qquad (2.12)
\end{aligned}
$$

$$
t^i = \left(\lambda e^m_m \delta^i_j + 2\mu e^i_j\right) n^j \qquad (2.13)
$$

$$
e^i_j = \frac{1}{2}\sqrt{\frac{g_{\overline{ii}}}{g_{\overline{jj}}}}\left[\frac{\partial}{\partial x^j}\left(\frac{u^i}{\sqrt{g_{\overline{ii}}}}\right) + \frac{g_{\overline{jj}}}{g_{\overline{ii}}}\frac{\partial}{\partial x^i}\left(\frac{u^j}{\sqrt{g_{\overline{jj}}}}\right) + \frac{1}{g_{\overline{ii}}}\sum_{m=1}^{3}\frac{\partial g_{\overline{ii}}}{\partial x^m}\frac{u^m}{\sqrt{g_{mm}}}\delta_{ij}\right] \qquad (2.14)
$$

The expression for n^j, the surface normal, is different for the three coordinate systems under study, thus the three normal surface tractions will be slightly different for each. Expressions for the surface normals and metric tensors can be inserted into e^i_j, the strain tensor, for the three coordinate systems and the indicated derivatives performed to arrive at the expressions given in the following sections. A simpler way to write the strain tensor e^i_j is with the introduction of the covariant partial derivative symbol ; and the Christoffel symbols of the second kind $\left\{ {i \atop mj} \right\}$. The Einstein summation convention is followed for all terms except those with bars over the index. In this case, the double indices are not "dummy" indices, but are fixed at the present value of the index.

$$e^i_j = \frac{1}{2}\sqrt{\frac{g_{\bar{i}\bar{i}}}{g_{\bar{j}\bar{j}}}}\left[\left(\frac{u^i}{\sqrt{g_{\bar{i}\bar{i}}}}\right)_{;j} + g_{nj}g^{im}\left(\frac{u^n}{\sqrt{g_{\bar{n}\bar{n}}}}\right)_{;m}\right]$$

$$\left(\frac{u^i}{\sqrt{g_{\bar{i}\bar{i}}}}\right)_{;j} = \frac{\partial}{\partial x^j}\left(\frac{u^i}{\sqrt{g_{\bar{i}\bar{i}}}}\right) + \left\{ {i \atop mj} \right\}\frac{u^m}{\sqrt{g_{\bar{m}\bar{m}}}} \tag{2.15}$$

Before we progress to the expressions for the displacements and normal surface tractions in the different coordinate systems, we must review some definitions for the different fields necessary to represent the entire solution. Let's begin by defining some commonly used terminology for the elastic scattering of waves. The different fields will be indicated by superscripts and subscripts. The incident field will be given an i superscript. The incident fields exist in the exterior field 1 only, thus all of the material parameters will have the subscript 1. The scattered field is also in field 1 and will also have a subscript 1, however the superscript will be lower case s. The fields in the interface layer, field 2, will be designated the refracted region via the common terminology from optics[49]. This field will be broken into fields moving towards the incident field and away from the incident field with the use of the tilde over functions when necessary to differentiate between the two fields. The tilde will indicate the field in the interface region propagating in the direction of the interior region. A subscript of 2 and lower case r superscript will be used in the interface region. Finally, the internal fields will be given a subscript 3 and a superscript t for transmitted, again following the common terminology for optics. These fields are graphically displayed in figure 2.2 for the different geometries. Since the two transverse waves travel at the same relative velocity in each medium, they are shown as a combined propagation vector in each region. In all cases, the use of a capitol (L,S,V) or (L,T) indicates the displacement field type.

In general we will have the twelve boundary conditions, six at the outer surface where the incident field is present,

$$
\begin{aligned}
u^j(incident) + u^j(scattered) &= u^j(refracted) \\
t^j(incident) + t^j(scattered) &= t^j(refracted)
\end{aligned}
\tag{2.16}
$$

and six at the inner boundary given by:

$$
\begin{aligned}
u^j(transmitted) &= u^j(refracted) \\
t^j(transmitted) &= t^j(refracted)
\end{aligned}
\tag{2.17}
$$

The refracted field is made up of two potentials and the vector component index j takes on the value of (x, y, z), (r, θ, z), or (r, θ, ϕ) for the components in each of the three coordinate systems. It should be remembered that the boundaries are a constant

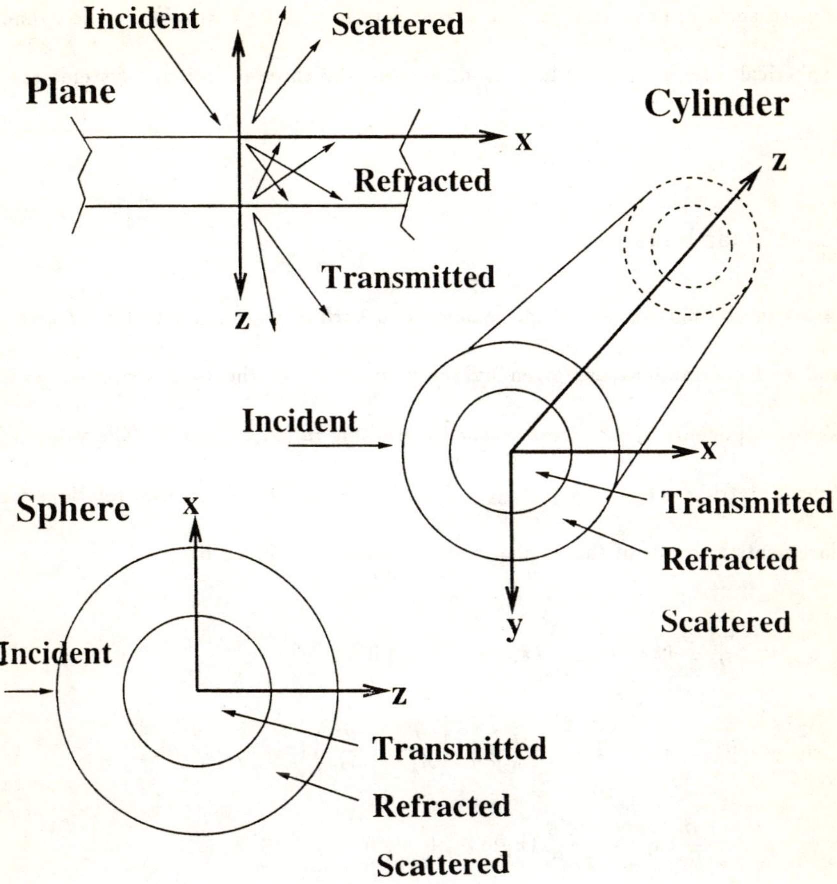

Figure 2.2: Field definitions for scattering from elastic interface layers.

coordinate surface, ($z = 0, b$) for the planar interface and ($r = a, b$) for the cylindrical and spherical interface. Boundary conditions for the three coordinate systems can now be given.

2.2.1 Planar boundary

In cartesian coordinates, the displacements and surface tractions simplify greatly. The normal surface tractions are given by the contraction of the base vector $\hat{e}_z$ with the physical components of the stress tensor t^i_j, resulting in $(\sigma_{xz}, \sigma_{yz}, \sigma_{zz})$. The values of the Euclidean metric are ($g_{11} = g_{22} = g_{33} = 1$) and $f_1 = 1$. The differential relations for the displacements in terms of the displacement potentials are given by:

$$
\begin{aligned}
u_x &= \frac{-i}{k}\frac{\partial}{\partial x}(\pi_L) + \frac{i}{K}\frac{\partial}{\partial z}(\pi_S) \quad + \quad 0 \\
u_y &= 0 \quad + \quad 0 \quad + \left(\frac{i}{K}\right)^2 \left(\frac{\partial^2}{\partial x^2} + \frac{\partial^2}{\partial z^2}\right)(\pi_V) \\
u_z &= \frac{-i}{k}\frac{\partial}{\partial z}(\pi_L) + \frac{-i}{K}\frac{\partial}{\partial x}(\pi_S) \quad + \quad 0
\end{aligned}
\tag{2.18}
$$

The traction components t^i_j can be written in the standard form σ_{ij} for cartesian coordinates since the physical components correspond to the stress components. Only three stresses with z components will be used in the boundary conditions, and they are given by:

$$
\begin{aligned}
\sigma_{zz} &= \frac{-i}{k}\left(\lambda\left(\frac{\partial^2}{\partial x^2} + \frac{\partial^2}{\partial z^2}\right) + 2\mu\frac{\partial^2}{\partial z^2}\right)(\pi_L) + 2\mu\left(\frac{-i}{K}\right)\frac{\partial^2}{\partial x \partial z}(\pi_S) + 0 \\
\sigma_{xz} &= 2\mu\left(\frac{-i}{k}\right)\frac{\partial^2}{\partial x \partial z}(\pi_L) + \mu\left(\frac{-i}{K}\right)\left(\frac{\partial^2}{\partial x^2} - \frac{\partial^2}{\partial z^2}\right)(\pi_S) + \quad 0
\end{aligned}
\tag{2.19}
$$

$$\sigma_{yz} \quad = \quad 0 \quad + \quad 0 \quad + \mu \left(\frac{i}{K}\right)^2 \frac{\partial}{\partial z} \left(\frac{\partial^2}{\partial x^2} + \frac{\partial^2}{\partial z^2}\right) (\pi_V)$$

If we look carefully at the displacements and stresses, one of the three fields (π_V) completely decouples from the problem. In the planar problem, the first potential or first curl potential π_S, is coupled to the longitudinal field. The same is true for the cylindrical problem for small angles, and in the spherical geometry it is the second curl potential π_V that couples with the longitudinal field. Many authors, in treating the planar problem use the propagation vector as the base vector for the transverse potentials rather than a fixed coordinate base vector. Since the propagation vector is in the plane of symmetry, the second transverse potential (π_V) becomes coupled to the longitudinal. In the present analysis, we will use the normal to the plane of symmetry and decouple the (π_V) field.

This decoupling reduces the problem to a system of four equations to solve for the four π_L and π_S fields for each interface, and a system of two equations for two π_V fields for each interface. The two interface problems will therefore lead to a system of eight equations and eight unknowns for the π_L and π_S potentials, and a system of four equations and four unknowns for the π_V potentials.

If we evaluate the displacements and normal surface tractions for the twelve different potential fields on the two different interfaces, we arrive at the following set of boundary conditions for π_L and π_S:

$(z = 0)$	$(z = d)$
$u_x^r + u_x^{\bar{r}} - u_x^s = u_x^i$	$u_x^r + u_x^{\bar{r}} - u_x^t = 0$

$$u_z^r + u_z^{\bar{r}} - u_z^s = u_z^i \qquad u_z^r + u_z^{\bar{r}} - u_z^t = 0$$

$$\sigma_{xz}^r + \sigma_{xz}^{\bar{r}} - \sigma_{xz}^s = \sigma_{xz}^i \qquad \sigma_{xz}^r + \sigma_{xz}^{\bar{r}} - \sigma_{xz}^t = 0$$

$$\sigma_{zz}^r + \sigma_{zz}^{\bar{r}} - \sigma_{zz}^s = \sigma_{zz}^i \qquad \sigma_{zz}^r + \sigma_{zz}^{\bar{r}} - \sigma_{zz}^t = 0$$

Similarly, the proper boundary conditions for π_V are as follows:

$$(z = 0) \qquad (z = d)$$

$$u_y^r + u_y^{\bar{r}} - u_y^s = u_y^i \qquad u_y^r + u_y^{\bar{r}} - u_y^t = 0$$

$$\sigma_{yz}^r + \sigma_{yz}^{\bar{r}} - \sigma_{yz}^s = \sigma_{yz}^i \qquad \sigma_{yz}^r + \sigma_{yz}^{\bar{r}} - \sigma_{yz}^t = 0$$

2.2.2 Cylindrical boundary

In cylindrical coordinates, the displacements and surface tractions also simplify greatly. The expressions are given below for the differential relations to the potentials of the displacements and those for the stresses in terms of the displacements. The expressions for the stresses in terms of the potentials can be found in Y.H. Pao and C.C. Mow's book [83] and are not repeated here for simplicity. The appropriate radial functions for the boundaries of the field covered by the potentials are Bessel and Hankel functions. The normal surface tractions are given by the contraction of the radial base vector $\hat{e}_r$ with the physical components of the stress tensor t_j^i resulting in $(\sigma_{rr}, \sigma_{r\theta}, \sigma_{rz})$. The values of the Euclidean metric are $(g_{11} = r^2, g_{22} = g_{33} = 1)$, with $f_1 = 1$ and $\vec{a} = \hat{e}_z$.

The displacements are given in terms of the potentials as:

$$u_r = \left(\frac{-i}{k}\right)\frac{\partial}{\partial r}(\pi_L) + \left(\frac{-i}{K}\right)\frac{1}{r}\frac{\partial}{\partial\theta}(\pi_S) + \left(\frac{i}{K}\right)^2\frac{\partial^2}{\partial r\partial z}(\pi_V)$$

$$u_\theta = \left(\frac{-i}{k}\right)\frac{1}{r}\frac{\partial}{\partial\theta}(\pi_L) - \left(\frac{-i}{K}\right)\frac{\partial}{\partial r}(\pi_S) - \left(\frac{i}{K}\right)^2\frac{1}{r}\frac{\partial^2}{\partial z\partial\theta}(\pi_V)$$

$$u_z = \left(\frac{-i}{k}\right)\frac{\partial}{\partial z}(\pi_L) + \quad 0 \quad + \left(\frac{i}{K}\right)^2\left(\frac{1}{r}\frac{\partial}{\partial r} + \frac{\partial^2}{\partial r^2} + \frac{\partial^2}{\partial\theta^2}\right)(\pi_V)$$

and the stresses (normal surface tractions) can be written in terms of the displacements as:

$$\sigma_{rr} = (\lambda + 2\mu)\frac{\partial u_r}{\partial r} + \lambda\left(\frac{1}{r}\frac{\partial u_\theta}{\partial\theta} + \frac{u_r}{r} + \frac{\partial u_z}{\partial z}\right)$$

$$\sigma_{r\theta} = \mu\left(\frac{1}{r}\frac{\partial u_r}{\partial\theta} + \frac{\partial u_\theta}{\partial r} - \frac{u_\theta}{r}\right)$$

$$\sigma_{rz} = \mu\left(\frac{\partial u_z}{\partial r} + \frac{\partial u_r}{\partial z}\right)$$

If we evaluate the displacements and normal surface tractions for the twelve different potential fields on the two different interfaces, we arrive at the following set of boundary conditions for π_L and π_S for small angles of incidence. We take the π_V field as decoupled from the π_L and π_S fields in this analysis resulting in:

$(r = b)$	$(r = a)$
$u_\theta^r + u_\theta^{\bar{r}} - u_\theta^s = u_\theta^i$	$u_\theta^r + u_\theta^{\bar{r}} - u_\theta^t = 0$
$u_r^r + u_r^{\bar{r}} - u_r^s = u_r^i$	$u_r^r + u_r^{\bar{r}} - u_r^t = 0$

$$\sigma^{r}_{r\theta} + \sigma^{\bar{r}}_{r\theta} - \sigma^{s}_{r\theta} = \sigma^{i}_{r\theta} \qquad \sigma^{r}_{r\theta} + \sigma^{\bar{r}}_{r\theta} - \sigma^{t}_{r\theta} = 0$$

$$\sigma^{r}_{rr} + \sigma^{\bar{r}}_{rr} - \sigma^{s}_{rr} = \sigma^{i}_{rr} \qquad \sigma^{r}_{rr} + \sigma^{\bar{r}}_{rr} - \sigma^{t}_{rr} = 0$$

The appropriate boundary conditions for the π_V field are:

$$(r = b) \qquad\qquad (r = a)$$

$$u^{r}_{z} + u^{\bar{r}}_{z} - u^{s}_{z} = u^{i}_{z} \qquad u^{r}_{z} + u^{\bar{r}}_{z} - u^{t}_{z} = 0$$

$$\sigma^{r}_{rz} + \sigma^{\bar{r}}_{rz} - \sigma^{s}_{rz} = \sigma^{i}_{rz} \qquad \sigma^{r}_{rz} + \sigma^{\bar{r}}_{rz} - \sigma^{t}_{rz} = 0$$

2.2.3 Spherical boundary

In spherical coordinates, the displacements and surface tractions can be written in terms of the potentials. The normal surface tractions are, as for cylindrical, given by the contraction of the radial base vector $\hat{e}_r$ with the physical components of the stress tensor t^i_j resulting in what we will call $(\sigma_{rr}, \sigma_{r\theta}, \sigma_{r\phi})$. The values of the Euclidean metric are $(g_{11} = 1, g_{22} = r^2, g_{33} = r^2 \sin^2\theta)$, $f_1 = r$, and $\vec{a} = r\hat{e}_r$.

The displacement can be written in terms of the potentials as:

$$\begin{aligned} u_r &= \left(\frac{-i}{k}\right)\left(\frac{\partial}{\partial r} - 1\right)(r\pi_L) \quad + \quad 0 \\ &+ \left(\frac{i}{K}\right)^2 \left(\frac{\cot\theta}{r^2}\frac{\partial}{\partial\theta} + \frac{\sin\theta}{r^2}\frac{\partial^2}{\partial\theta^2} + \frac{1}{r^2\sin^2\theta}\frac{\partial^2}{\partial\phi^2}\right)(r\pi_V) \qquad (2.20) \\ u_\theta &= \left(\frac{-i}{k}\right)\frac{1}{r^2}\frac{\partial}{\partial_\theta}(r\pi_L) + \left(\frac{-i}{K}\right)\frac{1}{r\sin\theta}\frac{\partial}{\partial_\phi}(r\pi_S) - \left(\frac{i}{K}\right)^2 \frac{1}{r}\frac{\partial^2}{\partial r\partial\theta}(r\pi_V) \end{aligned}$$

$$u_\phi = \left(\frac{-i}{k}\right)\frac{1}{r^2\sin\theta}\frac{\partial}{\partial_\phi}(r\pi_L) - \left(\frac{-i}{K}\right)\frac{1}{r}\frac{\partial}{\partial_\theta}(r\pi_S) - \left(\frac{i}{K}\right)^2\frac{1}{r\sin\theta}\frac{\partial^2}{\partial r\partial\phi}(r\pi_V)$$

and the stresses can then be written in terms of the displacements as:

$$\begin{aligned}
\sigma_{rr} &= \lambda\nabla\cdot\vec{u} + 2\mu\frac{\partial u_r}{\partial_r} \\
\sigma_{r\theta} &= \mu\left(\frac{\partial u_\theta}{\partial_r} - \frac{u_\theta}{r} + \frac{1}{r}\frac{\partial u_r}{\partial_\theta}\right) \\
\sigma_{r\phi} &= \mu\left(\frac{\partial u_\phi}{\partial_r} - \frac{u_\phi}{r} + \frac{1}{r\sin\theta}\frac{\partial u_r}{\partial_\theta}\right)
\end{aligned} \tag{2.21}$$

where

$$\nabla\cdot\vec{u} = \frac{1}{r^2}\frac{\partial}{\partial_r}\left(r^2 u_r\right) + \frac{1}{r\sin\theta}\frac{\partial}{\partial_\theta}(\sin\theta u_\theta) + \frac{1}{r\sin\theta}\frac{\partial}{\partial_\phi}(u_\phi)$$

If we evaluate the displacements and normal surface tractions for the twelve different potential fields on the two different interfaces, we arrive at the following set of boundary conditions for π_L and π_S:

$(r = b)$	$(r = a)$
$u_\theta^r + u_\theta^{\bar{r}} - u_\theta^s = u_\theta^i$	$u_\theta^r + u_\theta^{\bar{r}} - u_\theta^t = 0$
$u_r^r + u_r^{\bar{r}} - u_r^s = u_r^i$	$u_r^r + u_r^{\bar{r}} - u_r^t = 0$
$\sigma_{r\theta}^r + \sigma_{r\theta}^{\bar{r}} - \sigma_{r\theta}^s = \sigma_{r\theta}^i$	$\sigma_{r\theta}^r + \sigma_{r\theta}^{\bar{r}} - \sigma_{r\theta}^t = 0$
$\sigma_{rr}^r + \sigma_{rr}^{\bar{r}} - \sigma_{rr}^s = \sigma_{rr}^i$	$\sigma_{rr}^r + \sigma_{rr}^{\bar{r}} - \sigma_{rr}^t = 0$

The boundary conditions for π_V will be given by:

$$(r = b) \qquad\qquad (r = a)$$

$$u_\theta^r + u_\theta^{\bar{r}} - u_\theta^s = u_\theta^i \qquad\qquad u_\theta^r + u_\theta^{\bar{r}} - u_\theta^t = 0$$

$$\sigma_{r\theta}^r + \sigma_{r\theta}^{\bar{r}} - \sigma_{r\theta}^s = \sigma_{r\theta}^i \qquad\qquad \sigma_{r\theta}^r + \sigma_{r\theta}^{\bar{r}} - \sigma_{r\theta}^t = 0$$

2.3 Method of solution

In solving the algebraic system that results from evaluating the displacements and normal surface tractions at the two boundaries of the three regions, there will be a total of twelve unknown coefficients for the displacement potentials. When we allow for decoupling of one of the transverse fields, a maximum system of eight equations in eight unknowns results. Though generally considered intractable, we will algebraically solve the system of eight by eight equations for each of the problem geometries. To arrive at the solution to the system, and expressions for the coefficients of the displacement potentials, we will use Cramer's rule. The ratio of the determinants of the coefficient matrices will be equal to the coefficients, and the determinants will be reduced to tractable forms through extensive algebraic manipulations. In order to verify these long algebraic manipulations necessary to write the general solutions, we will resort to a machine verification of the analytic solution in two ways. The first will be to verify the forms of the analytic solution by symbolic-manipulation. This normally would be a task beyond the capabilities of most symbolic manipulators, since even with symbolic manipulators an 8×8 system of equations, where each element is itself a polynomial, is beyond the ability of present systems to handle independently. The knowledge of the correct way to manipulate the matrix can be used as a guide for the symbolic-manipulator which could not otherwise provide meaningful solutions.

We can therefore use the given formulas for displacements and normal surface tractions to write the boundary conditions as eight equations in the eight unknowns for an incident longitudinal or transverse plane wave of unit amplitude. For an incident spherical displacement wave, generated from a specific point in the field, the expressions are exactly analogous but with slightly altered expressions for the incident fields and displacement potentials [47].

In this section we discuss the proper selection of potentials for the three different types of fields in each of the different geometries. For the modal coefficients of the potentials, we will use lower case (a, b, c) to indicate the coefficients produced by an incident longitudinal field and upper case (A, B, C) for an incident transverse field. A subscript and tilde will also be used to indicate the fields by number and type on the coefficients. The tilde will indicate the field in the interface region propagating in the direction of the interior. Thus in general we will have twelve potential fields, but for each case only a maximum of eight will be coupled together.

2.3.1 Planar-cartesian coordinates

The specific expressions for the displacement potentials in the case of an incident planar displacement wave on the planar interface between two elastic half spaces can be readily expressed in terms of the wavenumbers and angles of propagation relative to the primary propagation direction. The planar displacement fields are single term solutions to the

Helmholtz equation and are given as follows for incident fields:

$$\pi_L^i = e^{i\vec{k}_1\cdot\vec{z}}e^{i\vec{k}_1\cdot\vec{x}}, \quad \pi_S^i = e^{i\vec{K}_1\cdot\vec{z}}e^{i\vec{K}_1\cdot\vec{x}}, \quad \pi_V^i = e^{i\vec{K}_1\cdot\vec{z}}e^{i\vec{K}_1\cdot\vec{x}} \tag{2.22}$$

scattered fields:

$$\pi_L^s = \begin{bmatrix} a_1 \\ A_1 \end{bmatrix} e^{i\vec{k}_1\cdot\vec{z}}e^{i\vec{k}_1\cdot\vec{x}}, \quad \pi_S^s = \begin{bmatrix} b_1 \\ B_1 \end{bmatrix} e^{i\vec{K}_1\cdot\vec{z}}e^{i\vec{K}_1\cdot\vec{x}}, \quad \pi_V^s = \begin{bmatrix} c_1 \\ C_1 \end{bmatrix} e^{i\vec{K}_1\cdot\vec{z}}e^{i\vec{K}_1\cdot\vec{x}}$$

refracted fields:

$$\pi_L^r = \begin{bmatrix} a_2 \\ A_2 \end{bmatrix} e^{i\vec{k}_2\cdot\vec{z}}e^{i\vec{k}_2\cdot\vec{x}}, \quad \pi_S^r = \begin{bmatrix} b_2 \\ B_2 \end{bmatrix} e^{i\vec{K}_2\cdot\vec{z}}e^{i\vec{K}_2\cdot\vec{x}}, \quad \pi_V^r = \begin{bmatrix} c_2 \\ C_2 \end{bmatrix} e^{i\vec{K}_2\cdot\vec{z}}e^{i\vec{K}_2\cdot\vec{x}}$$

$$\tilde{\pi}_L^r = \begin{bmatrix} \tilde{a}_2 \\ \tilde{A}_2 \end{bmatrix} e^{i\vec{k}_2\cdot\vec{z}}e^{i\vec{k}_2\cdot\vec{x}}, \quad \tilde{\pi}_S^r = \begin{bmatrix} \tilde{b}_2 \\ \tilde{B}_2 \end{bmatrix} e^{i\vec{K}_2\cdot\vec{z}}e^{i\vec{K}_2\cdot\vec{x}}, \quad \tilde{\pi}_V^r = \begin{bmatrix} \tilde{c}_2 \\ \tilde{C}_2 \end{bmatrix} e^{i\vec{K}_2\cdot\vec{z}}e^{i\vec{K}_2\cdot\vec{x}}$$

and transmitted fields:

$$\pi_L^t = \begin{bmatrix} a_3 \\ A_3 \end{bmatrix} e^{i\vec{k}_3\cdot\vec{z}}e^{i\vec{k}_3\cdot\vec{x}}, \quad \pi_S^t = \begin{bmatrix} b_3 \\ B_3 \end{bmatrix} e^{i\vec{K}_3\cdot\vec{z}}e^{i\vec{K}_3\cdot\vec{x}}, \quad \pi_V^t = \begin{bmatrix} c_3 \\ C_3 \end{bmatrix} e^{i\vec{K}_3\cdot\vec{z}}e^{i\vec{K}_3\cdot\vec{x}}$$

2.3.2 Cylindrical coordinates

In cylindrical coordinates, we follow the approach of Y.H. Pao and C.C. Mow [83] in writing the displacement potentials. The displacement potentials are composed of Fourier Bessel series with an additional term i^n added to simplify the later algebra. The additional terms ϵ_n, and i^n show up in the decomposition of the incident planar displacement fields into incident cylindrical displacement fields. The displacement potentials for the incident planar displacement field and the scattered, refracted (in the shell), and transmitted displacement fields are given as follows for incident fields:

$$\pi_L^i = \sum_{n=0}^{\infty} \epsilon_n i^n J_n(\vec{k}_1 \cdot \vec{r}) e^{in\theta} e^{i\vec{k}_1 \cdot \vec{z}}$$

$$\pi_S^i = \sum_{n=0}^{\infty} \epsilon_n i^n J_n(\vec{K}_1 \cdot \vec{r}) e^{in\theta} e^{i\vec{K}_1 \cdot \vec{z}} \qquad \epsilon_n = 1(n=0), \epsilon_n = 2(n \neq 0)$$

$$\pi_V^i = \sum_{n=0}^{\infty} \epsilon_n i^n J_n(\vec{K}_1 \cdot \vec{r}) e^{in\theta} e^{i\vec{K}_1 \cdot \vec{z}} \tag{2.23}$$

scattered fields:

$$\pi_L^s = \sum_{n=0}^{\infty} \begin{bmatrix} a_{1n} \\ A_{1n} \end{bmatrix} i^n H_n^{(1)}(\vec{k}_1 \cdot \vec{r}) e^{in\theta} e^{i\vec{k}_1 \cdot \vec{z}}$$

$$\pi_S^s = \sum_{n=0}^{\infty} \begin{bmatrix} b_{1n} \\ B_{1n} \end{bmatrix} i^n H_n^{(1)}(\vec{K}_1 \cdot \vec{r}) e^{in\theta} e^{i\vec{K}_1 \cdot \vec{z}}$$

$$\pi_V^s = \sum_{n=0}^{\infty} \begin{bmatrix} c_{1n} \\ C_{1n} \end{bmatrix} i^n H_n^{(1)}(\vec{K}_1 \cdot \vec{r}) e^{in\theta} e^{i\vec{K}_1 \cdot \vec{z}} \tag{2.24}$$

refracted fields:

$$\pi_L^r = \sum_{n=0}^{\infty} \begin{bmatrix} a_{2n} \\ A_{2n} \end{bmatrix} i^n H_n^{(1)}(\vec{k}_2 \cdot \vec{r}) e^{in\theta} e^{i\vec{k}_2 \cdot \vec{z}}$$

$$\pi_S^r = \sum_{n=0}^{\infty} \begin{bmatrix} b_{2n} \\ B_{2n} \end{bmatrix} i^n H_n^{(1)}(\vec{K}_2 \cdot \vec{r}) e^{in\theta} e^{i\vec{K}_2 \cdot \vec{z}}$$

$$\pi_V^r = \sum_{n=0}^{\infty} \begin{bmatrix} c_{2n} \\ C_{2n} \end{bmatrix} i^n H_n^{(1)}(\vec{K}_2 \cdot \vec{r}) e^{in\theta} e^{i\vec{K}_2 \cdot \vec{z}}$$

$$\tilde{\pi}_L^r = \sum_{n=0}^{\infty} \begin{bmatrix} \tilde{a}_{2n} \\ \tilde{A}_{2n} \end{bmatrix} i^n H_n^{(2)}(\vec{k}_2 \cdot \vec{r}) e^{in\theta} e^{i\vec{k}_2 \cdot \vec{z}}$$

$$\tilde{\pi}_S^r = \sum_{n=0}^{\infty} \begin{bmatrix} \tilde{b}_{2n} \\ \tilde{B}_{2n} \end{bmatrix} i^n H_n^{(2)}(\vec{K}_2 \cdot \vec{r}) e^{in\theta} e^{i\vec{K}_2 \cdot \vec{z}}$$

$$\tilde{\pi}_V^r = \sum_{n=0}^{\infty} \begin{bmatrix} \tilde{c}_{2n} \\ \tilde{C}_{2n} \end{bmatrix} i^n H_n^{(2)}(\vec{K}_2 \cdot \vec{r}) e^{in\theta} e^{i\vec{K}_2 \cdot \vec{z}} \tag{2.25}$$

and transmitted fields:

$$\pi_L^t = \sum_{n=0}^{\infty} \begin{bmatrix} a_{3n} \\ A_{3n} \end{bmatrix} i^n J_n(\vec{k}_3 \cdot \vec{r}) e^{in\theta} e^{i\vec{k}_3 \cdot \vec{z}}$$

$$\pi_S^t = \sum_{n=0}^{\infty} \begin{bmatrix} b_{3n} \\ B_{3n} \end{bmatrix} i^n J_n(\vec{K}_3 \cdot \vec{r}) e^{in\theta} e^{i\vec{K}_3 \cdot \vec{z}}$$

$$\pi_V^t = \sum_{n=0}^{\infty} \begin{bmatrix} c_{3n} \\ C_{3n} \end{bmatrix} i^n J_n(\vec{K}_3 \cdot \vec{r}) e^{in\theta} e^{i\vec{K}_3 \cdot \vec{z}} \tag{2.26}$$

2.3.3 Spherical coordinates

In spherical coordinates, the displacement potentials for the three separate fields are written with two subscripts, one to indicate the exterior (1), shell (2), and interior (3), the other to indicate longitudinal (L), uncoupled transverse (S), coupled transverse (V). In the literature these fields are given the identifications L-longitudinal, SH-uncoupled transverse, and SV-coupled transverse. Normalization factors show up in the incident field expansions that, as in cylindrical coordinates, could also multiply the modal coefficients to simplify the later algebra. The incident fields are given by:

$$k_1 r \pi_L^i = \sum_{l=0}^{\infty} i^{l+1} (2l+1)\, \psi_l(k_1 r) P_l(\cos\theta)$$

$$K_1 r \pi_S^i = \sum_{l=1}^{\infty} i^l \left(\frac{2l+1}{l(l+1)} \right) \psi_l(K_1 r) P_l^{(1)}(\cos\theta) \sin\phi$$

$$K_1 r \pi_V^i = \sum_{l=1}^{\infty} i^{l-1} \left(\frac{2l+1}{l(l+1)} \right) \psi_l(K_1 r) P_l^{(1)}(\cos\theta) \cos\phi \tag{2.27}$$

with the scattered fields:

$$k_1 r \pi_L^s = \sum_{l=0}^{\infty} \left[\begin{array}{c} a_{1l} \\ A_{1l} \end{array} \right] i^{l+1} \zeta_l^{(1)}(k_1 r) P_l(\cos\theta)$$

$$K_1 r \pi_S^s = \sum_{l=1}^{\infty} \left[\begin{array}{c} b_{1l} \\ B_{1l} \end{array} \right] i^{l} \zeta_l^{(1)}(K_1 r) P_l^{(1)}(\cos\theta) \sin\phi$$

$$K_1 r \pi_V^s = \sum_{l=1}^{\infty} \left[\begin{array}{c} c_{1l} \\ C_{1l} \end{array} \right] i^{l-1} \zeta_l^{(1)}(K_1 r) P_l^{(1)}(\cos\theta) \cos\phi \tag{2.28}$$

refracted fields:

$$k_2 r \pi_L^r = \sum_{l=0}^{\infty} \left[\begin{array}{c} a_{2l} \\ A_{2l} \end{array} \right] i^{l+1} \zeta_l^{(1)}(k_2 r) P_l(\cos\theta)$$

$$K_2 r \pi_S^r = \sum_{l=1}^{\infty} \left[\begin{array}{c} b_{2l} \\ B_{2l} \end{array} \right] i^{l} \zeta_l^{(1)}(K_2 r) P_l^{(1)}(\cos\theta) \sin\phi$$

$$K_2 r \pi_V^r = \sum_{l=1}^{\infty} \left[\begin{array}{c} c_{2l} \\ C_{2l} \end{array} \right] i^{l-1} \zeta_l^{(1)}(K_2 r) P_l^{(1)}(\cos\theta) \cos\phi$$

$$k_2 r \tilde{\pi}_L^r = \sum_{l=0}^{\infty} \left[\begin{array}{c} \tilde{a}_{2l} \\ \tilde{A}_{2l} \end{array} \right] i^{l+1} \zeta_l^{(2)}(k_2 r) P_l(\cos\theta)$$

$$K_2 r \tilde{\pi}_S^r = \sum_{l=1}^{\infty} \left[\begin{array}{c} \tilde{b}_{2l} \\ \tilde{B}_{2l} \end{array} \right] i^{l} \zeta_l^{(2)}(K_2 r) P_l^{(1)}(\cos\theta) \sin\phi$$

$$K_2 r \tilde{\pi}_V^r = \sum_{l=1}^{\infty} \left[\begin{array}{c} \tilde{c}_{2l} \\ \tilde{C}_{2l} \end{array} \right] i^{l-1} \zeta_l^{(2)}(K_2 r) P_l^{(1)}(\cos\theta) \cos\phi \tag{2.29}$$

and transmitted fields:

$$k_3 r \pi_L^t = \sum_{l=0}^{\infty} \begin{bmatrix} a_{3l} \\ A_{3l} \end{bmatrix} i^{l+1} \psi_l(k_3 r) P_l(\cos\theta)$$

$$K_3 r \pi_S^t = \sum_{l=1}^{\infty} \begin{bmatrix} b_{3l} \\ B_{3l} \end{bmatrix} i^{l} \psi_l(K_3 r) P_l^{(1)}(\cos\theta)\sin\phi$$

$$K_3 r \pi_V^t = \sum_{l=1}^{\infty} \begin{bmatrix} c_{3l} \\ C_{3l} \end{bmatrix} i^{l-1} \psi_l(K_3 r) P_l^{(1)}(\cos\theta)\cos\phi \tag{2.30}$$

2.4 Incident field expansions

In order to study the scattering of elastic waves with the method of Clebsch and Mie, an expansion of the incident planar displacements waves in terms of the appropriate displacement potentials of the problem geometry must be derived. The incident fields will be either compressional (longitudinal, dilatational) or shear (transverse, distortional) waves, and although we discuss only incident plane waves here, incident spherical waves can be similarly expanded in the appropriate potential-functions and solved in the same manner [47].

2.4.1 Plane wave expansions

In solving the planar interface problem, the incident fields are given in the appropriate form for immediate substitution into the boundary conditions. These fields can be taken as either longitudinal or transverse, with the same form as previously stated for the potentials given by:

$$\pi_L^i = e^{i\vec{k}_1\cdot\vec{z}} e^{i\vec{k}_1\cdot\vec{x}}, \quad \pi_S^i = e^{i\vec{K}_1\cdot\vec{z}} e^{i\vec{K}_1\cdot\vec{x}}, \quad \pi_V^i = e^{i\vec{K}_1\cdot\vec{z}} e^{i\vec{K}_1\cdot\vec{x}} \tag{2.31}$$

2.4.2 Cylindrical wave expansions

In cylindrical coordinates, the expansion of the incident plane waves follows the approach of Y.H. Pao and C.C. Mow [83]. Incident longitudinal and transverse fields can be expanded in the same manner so we will begin with a detailed derivation of the longitudinal potential and finish with a simplified derivation of the transverse potential. We begin by writing the incident plane wave propagating in, and with displacements in, the $\hat{x}$ direction. The displacement vector is

$$\vec{u}_L = \hat{x}e^{ikx} = \frac{-i}{k}\nabla\pi_L^i \tag{2.32}$$

Equating the $\hat{x}$ components

$$\{\vec{u}_L\}_{\hat{x}} = \hat{x}e^{ikx} = \frac{-i}{k}\frac{\partial}{\partial x}\pi_L^i \tag{2.33}$$

and integrating with respect to x and solving for π_L^i gives

$$\pi^i = e^{ikx} \tag{2.34}$$

Similarly, we can derive the appropriate potential for a propagation vector with $\hat{x}$ and $\hat{z}$ components as

$$\pi^i = e^{i\vec{k}\cdot\vec{x}}e^{i\vec{k}\cdot\vec{z}}e^{-i\omega t} \tag{2.35}$$

where in cylindrical coordinates $x = r\cos\theta$ and we let $\vec{k}\cdot\vec{x} = k_x x$. Then the resulting factor $e^{ik_x r\cos\theta}$ can be written as a complex Fourier series expansion

$$e^{ik_x r\cos\theta} = \sum_{n=-\infty}^{+\infty} C_n(r)e^{in\theta} \tag{2.36}$$

Solving for the coefficients $C_n(r)$, we can write

$$C_n(r) = \frac{1}{2\pi}\int_0^{2\pi} e^{ik_x r\cos\theta}e^{-in\theta}d\theta \tag{2.37}$$

If $e^{-in\theta}$ is broken into $\cos n\theta - i\sin n\theta$ then the $\sin n\theta$ portion of the integral is zero and the $\cos n\theta$ portion is

$$C_n(r) = \frac{1}{\pi}\int_0^{\pi} e^{ik_x r\cos\theta}\cos n\theta d\theta = i^n J_n(k_x r) \tag{2.38}$$

Then substituting back in for the complex Fourier series expansion gives

$$e^{ik_x r\cos\theta} = \sum_{n=-\infty}^{+\infty} i^n J_n(k_x r)e^{in\theta} \tag{2.39}$$

Breaking into parts and symmetrizing the summation gives

$$e^{ik_x r\cos\theta} = J_0(k_x r) + \sum_{n=0}^{+\infty} i^n J_n(k_x r)\left(e^{in\theta} + e^{-in\theta}\right) \tag{2.40}$$

Defining $\epsilon_0 = 1, \epsilon_n = 2(n > 0)$, we can then write

$$e^{ik_x r\cos\theta} = \sum_{n=0}^{+\infty} \epsilon_n i^n J_n(k_x r)\cos n\theta \tag{2.41}$$

and finally substituting back into the expression for the potential to give

$$\pi^i = \sum_{n=0}^{+\infty} \epsilon_n i^n J_n(\vec{k}\cdot\hat{x}r)e^{in\theta}e^{i\vec{k}\cdot\vec{z}}e^{-i\omega t} \tag{2.42}$$

where $\cos n\theta$ is rewritten in exponential form and the time dependence is usually excluded from the potentials.

For an incident transverse field, we take the displacements in the $-\hat{y}$ direction, given by

$$\vec{u}_S = -\hat{y}e^{iKx} \quad = -\left(\hat{e}_r\sin\theta + \hat{e}_\theta\cos\theta\right)e^{iKr cos\theta} \tag{2.43}$$

The displacements due to the transverse potential field can be written as

$$\vec{u}_S = \frac{-i}{K}\nabla\times(\hat{z}\pi_S) \quad = \frac{-i}{K}\left(\frac{\hat{e}_r}{r}\frac{\partial}{\partial\theta} - \hat{e}_\theta\frac{\partial}{\partial r}\right)(\pi_S) \tag{2.44}$$

Now if we equate the radial component of the incident field with the radial component of the transverse potential field π_S^i we arrive at

$$\frac{\partial}{\partial\theta}(\pi_S^i) = -iKr\sin\theta e^{iKr\cos\theta} \tag{2.45}$$

We notice that the expression on the right is a derivative with respect to θ, thus we can integrate and set the arbitrary constant to zero, giving for the potential, the same expression as the longitudinal potential. After adding the z and t dependence which was previously suppressed for simplicity we arrive at the following expression for the incident transverse displacement potential

$$\begin{aligned} \pi_S^i &= e^{iKr\cos\theta} \\ \pi_S^i &= \sum_{n=0}^{+\infty} \epsilon_n i^n J_n(\vec{k}\cdot\hat{x}r)e^{in\theta}e^{i\vec{k}\cdot\vec{z}}e^{-i\omega t} \end{aligned} \tag{2.46}$$

2.4.3 Spherical wave expansions

Expansion of the incident plane waves in spherical coordinates is similar in approach to the expansion in cylindrical coordinates with the use of Bauer's formula

$$e^{ikr\cos\theta} = \sum_{l=0}^{\infty} i^l(2l+1)\frac{\psi_l(kr)}{(kr)}P_l(\cos\theta) \tag{2.47}$$

and some relations for associated Legendre functions to derive the incident field expansions which are given by:

$$\begin{aligned} (kr)\pi_L^i &= \sum_{l=0}^{\infty} i^{l+1}(2l+1)\psi_l(k_1r)P_l(\cos\theta) \\ (Kr)\pi_S^i &= \sum_{l=1}^{\infty} i^l\left(\frac{2l+1}{l(l+1)}\right)\psi_l(K_1r)P_l^{(1)}(\cos\theta)\sin\phi \end{aligned}$$

$$(Kr)\pi_V^i = \sum_{l=1}^{\infty} i^{l-1}\left(\frac{2l+1}{l(l+1)}\right)\psi_l(K_1 r)P_l^{(1)}(\cos\theta)\cos\phi \tag{2.48}$$

We will first consider the incident field expansion of the longitudinal potential and then briefly discuss the extension to the transverse fields. We take the incident, unit amplitude, plane wave with displacement and propagation in the z-direction. Equating the longitudinal potential to the plane wave displacement vector gives

$$\vec{u}_L = \hat{z}e^{ikz} = \frac{-i}{k}\nabla\pi_L^i \tag{2.49}$$

Transforming to spherical coordinates results in:

$$\begin{aligned}\vec{u}_L &= (\hat{e}_r\cos\theta - \hat{e}_\theta\sin\theta)\,e^{ikr\cos\theta} \\ &= \frac{-i}{k}\left(\hat{e}_r\frac{\partial}{\partial r} + \frac{\hat{e}_\theta}{r}\frac{\partial}{\partial\theta} + \frac{\hat{e}_\phi}{r\sin\theta}\frac{\partial}{\partial\phi}\right)(\pi_L^i)\end{aligned} \tag{2.50}$$

Now we equate the radial components of the two equations

$$\frac{-i}{k}\frac{\partial}{\partial r}(\pi_L^i) = \cos\theta e^{ikr\cos\theta} \tag{2.51}$$

We notice that the right hand side is a derivative with respect to r and integrate, setting the integration constant to zero and solving for the longitudinal potential

$$\pi_L^i = e^{ikr\cos\theta} \tag{2.52}$$

Using Bauer's formula we arrive at the previously given longitudinal potential

$$\pi_L^i = \sum_{l=0}^{\infty} i^l(2l+1)\frac{\psi_l(kr)}{kr}P_l(\cos\theta) \tag{2.53}$$

Similarly, the second transverse field π_V can be written in terms of an incident displacement field as

$$\vec{u}_V = \hat{x}e^{iKz} = -\left(\frac{i}{K}\right)^2\nabla\times\nabla\times\left(\hat{e}_r Kr\pi_V^i\right) \tag{2.54}$$

Changing to spherical coordinates and equating radial components

$$
\begin{aligned}
\{\vec{u}_V\}_{\hat{r}} &= \sin\theta\cos\phi e^{iKr\cos\theta} \\
&= -\left(\frac{i}{K}\right)^2\left(\frac{\cot\theta}{r^2}\frac{\partial}{\partial\theta}+\frac{1}{r^2}\frac{\partial^2}{\partial\theta^2}+\frac{1}{r^2\sin^2\theta}\frac{\partial^2}{\partial\phi^2}\right)\left(Kr\pi_V^i\right) \qquad (2.55)
\end{aligned}
$$

If we use the radial and angular separation equations, and notice that the planar displacement field can be written as the derivative with respect to θ we arrive at

$$
-\frac{\partial}{\partial\theta}\frac{1}{iKr}e^{iKr\cos\theta}\cos\phi = -\left(\frac{i}{K}\right)^2\frac{l(l+1)}{r^2}\left(Kr\pi_V^i\right) \qquad (2.56)
$$

We can now expand the exponential function using Bauer's formula and introduce the relation for Legendre functions $-\frac{\partial}{\partial\theta}P_l(\cos\theta) = P_l^{(1)}(\cos\theta)$ to give

$$
\sum_{l=1}^{\infty} i^{l-1}(2l+1)\frac{\psi_l(Kr)}{Kr}P_l^{(1)}(\cos\theta)\cos\phi = \frac{1}{Kr}l(l+1)\left(Kr\pi_V^i\right) \qquad (2.57)
$$

Solving for (π_V^i) gives the following expression for the incident transverse field

$$
(Kr)\pi_V^i = \sum_{l=1}^{\infty} i^{l-1}\frac{(2l+1)}{l(l+1)}\psi_l(Kr)P_l^{(1)}(\cos\theta)\cos\phi \qquad (2.58)
$$

The uncoupled incident transverse field (π_S) can be expanded in the same manner and is not presented here for simplicity.

Chapter 3

PLANAR LAYERED INTERFACE

3.1 Introduction

In the analysis of incident elastic displacement waves reflecting from a planar interface separating two elastic half spaces, we are primarily interested in gaining insight to apply to the cylindrical and spherical geometries. While many authors have studied the planar layered interface problem, with the intent of studying seismic wave propagation and transmission modes of laminated composites, their focus has been on a slightly different elastic scattering problem. Since we are interested in modeling the interaction between a laminated composite and the reflected fields due to an incident planar displacement field, we must solve the full system of equations to arrive at expressions for the scattered fields. An algebraic solution to the resulting 8×8 system of boundary conditions does not appear in the literature, since it is generally considered too complicated for algebraic solutions and numerical solutions can be readily computed.

In this section, we will review the formal solution to the elastic wave scattering from a planar interface separating two elastic half spaces, providing the analytic expressions for the scattered displacement fields. Three different incident displacement fields will be considered as incident on the interface layer originating in the exterior field. The incident fields are: a longitudinal displacement field, a coupled transverse displacement field, and an uncoupled transverse displacement field. In the planar case, we have taken $\vec{a} = \hat{y}$ in computing the curls of the transverse displacement potentials. We could just as easily have taken any of the base vectors, or even a linear combination of the base vectors representing the three dimensional propagation vector. The geometry of the elastic scattering problem is given by figure 3.1 for reference.

In order to set up the appropriate boundary value problem for the elastic wave scattering from an interface layer, we first evaluate the relationships between the displacements, normal surface tractions, and displacement potentials. We will begin with general expressions that can be evaluated in the different fields to meet the boundary conditions then specialize the relations for the specific regions. This is a necessary step for such a lengthy problem. Moving directly from the differential relations to the field evaluated expressions at the boundaries introduces a source of possible error that is avoided by this "symbolic" first step.

An additional $\left(\pm\right)$ is included in the expressions to distinguish between waves traveling in the positive and negative $\hat{z}$ direction. The value of γ is determined by the direction of propagation and is the projection of the wavenumber times propagation vector on the $\hat{x}$ axis, perpendicular to the primary direction of propagation and constant for all fields. The displacements can be written in terms of the potentials as:

$$
\begin{aligned}
u_x &= \left(\frac{\gamma}{k}\right)\pi_L + \quad 0 \quad - \left(\frac{\pm K\cos\phi^V}{K}\right)\pi_S \\
u_y &= 0 \quad + \quad \pi_V \quad + \quad 0 \\
u_z &= \left(\frac{\pm k\cos\phi^L}{k}\right)\pi_L + \quad 0 \quad + \quad \left(\frac{\gamma}{K}\right)\pi_S
\end{aligned}
\tag{3.1}
$$

We see immediately that the displacements, regardless of the field, couple only the longitudinal (π_L) and (π_S) displacement potentials. The components t^i_j can also be writ-

Planar Interface Layer

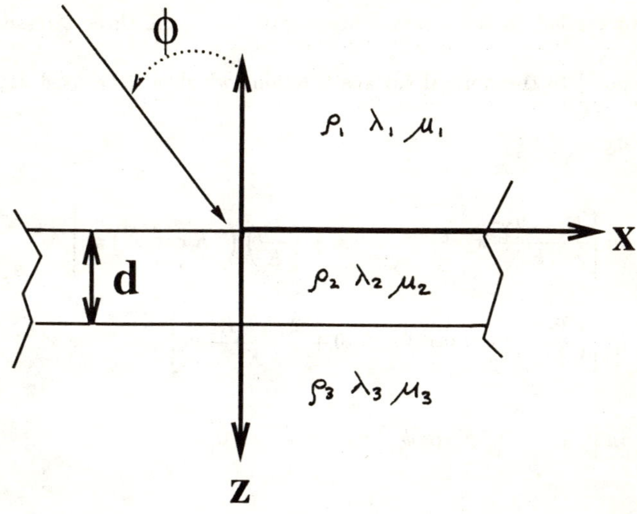

Figure 3.1: Problem geometry for the planar interface layer.

ten in terms of the displacement potentials in this symbolic form. We use the standard notation for stress σ_{ij} for cartesian coordinates since in cartesian coordinates the traction components correspond to the stress components. Only the three stresses with $\hat{z}$ components correspond to the normal surface tractions which will be used as the boundary conditions giving:

$$\begin{aligned}
\sigma_{zz} &= i\mu\left[\frac{K^2-2\gamma^2}{k}\pi_L + \quad 0 \quad + \left(\frac{2\gamma}{K}\right)\left(\pm K\cos\phi^V\right)\pi_S\right] \\
\sigma_{xz} &= i\mu\left[\left(\frac{2\gamma}{k}\right)\left(\pm k\cos\phi^L\right)\pi_L + 0 + \frac{2\gamma^2-K^2}{K}\pi_S\right] \\
\sigma_{yz} &= i\mu\left[0 + \quad \left(\pm K\cos\phi^V\right)\pi_S \quad + \quad 0\right]
\end{aligned} \tag{3.2}$$

The proper boundary conditions are the continuity of displacement and normal surface traction given above, evaluated for each field at each of the planar interfaces shown in figure 3.1. This gives six boundary conditions per interface, but as we can see from the displacements and stresses, four decouple from the six. This means that the π_V wave cannot be excited by an incident π_L or π_S wave, reducing the problem to a system of four equations and a system of two equations for each interface. The two interface problem will therefore lead to a system of eight equations and eight unknowns for the π_L and π_S potentials, and a system of four equations and four unknowns for the π_V potentials.

We can write the system of boundary value equations symbolically as $Mx = b$ for the longitudinal displacement potentials and coupled transverse displacement potentials. In this equation, M is the matrix of boundary conditions constituting the displacements and normal surface tractions on the two boundaries. In addition, b is the resulting column of displacements and normal surface tractions due to the incident longitudinal or transverse wave. For the uncoupled displacement potentials we can write $Nx = c$ in the same manner as the coupled problem, where N is the boundary condition matrix and c are the boundary conditions on the exterior surface of the interface due to the incident transverse displacement wave. The algebraic solutions to these problems can then be written in terms of the ratio of the determinants of the coefficient matricies via Cramer's rule.

3.2 Incident longitudinal wave

If we have an incident longitudinal (compressional-ultrasonic) wave of unit amplitude traveling primarily in the $\hat{z}$ direction, scattered, refracted, and transmitted displacement fields will be generated of both longitudinal and transverse types (see figure 3.1). We begin by writing the incident planar displacement wave as a displacement potential

$$\pi_L^i = e^{ik_1 \cos\phi^i z} e^{ik_1 \sin\phi^i x} \tag{3.3}$$

The eight unknown coefficients of the displacement potentials, defined previously, can be solved by evaluating the boundary conditions and solving the resulting algebraic system by Cramer's rule. We substitute in for the propagation vector of each of the displacement fields via Snell's law. The angles given by ϕ^i are defined as the acute angles to the $\hat{z}$ axis. We need not distinguish between the two refracted field angles, since in this analysis they are always equal in order to satisfy Snell's law of refraction. This is also true of the scattered field of the same type as the incident. In this case, the scattered longitudinal field propagates at the same acute angle with respect to the $\hat{z}$ axis as the incident longitudinal field but in the opposite direction, $\phi_L^s = \phi^i$. We can now write the appropriate displacement potentials for the planar problem as:

$$
\begin{aligned}
\pi_L^s &= a_1 e^{ik_1 \cos\phi_L^s z} e^{ik_1 \sin\phi_L^s x}, \quad \pi_S^s = b_1 e^{iK_1 \cos\phi_S^s z} e^{iK_1 \sin\phi_S^s x} \\
\pi_L^r &= a_2 e^{ik_2 \cos\phi_L^r z} e^{ik_2 \sin\phi_L^r x}, \quad \pi_S^r = b_2 e^{iK_2 \cos\phi_S^r z} e^{iK_2 \sin\phi_S^r x} \\
\tilde{\pi}_L^r &= \tilde{a}_2 e^{ik_2 \cos\phi_L^r z} e^{ik_2 \sin\phi_L^r x}, \quad \tilde{\pi}_S^r = \tilde{b}_2 e^{iK_2 \cos\phi_S^r z} e^{iK_2 \sin\phi_S^r x}, \\
\pi_L^t &= a_3 e^{ik_3 \cos\phi_L^t z} e^{ik_3 \sin\phi_L^t x}, \quad \pi_S^t = b_3 e^{iK_3 \cos\phi_S^t z} e^{iK_3 \sin\phi_S^t x} \qquad (3.4)
\end{aligned}
$$

By applying the boundary conditions, as discussed in detail in chapter two, we arrive at eight equations for displacements and normal surface tractions that must be continuous across the two boundaries. The eight unknown coefficients of the displacement potentials can then be solved for. Filling the matrix M with the following set of terms, we can write the most general form for bounding planes at $z = z_1$ and $z = z_2$ ($z_2 - z_1 = d$). Usually, $z_1 = 0$ for convenience and the exponential functions in the upper half of the matrix M can be set to unity. The elements of M are:

$$
\begin{aligned}
M[1,1] &= -\left(\frac{\gamma}{k_1}\right) \\
M[1,2] &= -\left(\frac{K_1 \cos\phi_S^s}{K_1}\right) \\
M[1,3] &= \left(\frac{\gamma}{k_2}\right) e^{+ik_2 \cos\phi_L^r z_1} \\
M[1,4] &= \left(\frac{\gamma}{k_2}\right) e^{-ik_2 \cos\phi_L^r z_1} \\
M[1,5] &= -\left(\frac{K_2 \cos\phi_S^r}{K_2}\right) e^{+iK_2 \cos\phi_S^r z_1}
\end{aligned}
$$

$$M[1,6] = \left(\frac{K_2 \cos\phi_S^r}{K_2}\right) e^{-iK_2 \cos\phi_S^r z_1}$$

$$M[1,7] = M[1,8] = 0$$

$$M[2,1] = \left(\frac{k_1 \cos\phi_L^s}{k_1}\right)$$

$$M[2,2] = -\left(\frac{\gamma}{K_1}\right)$$

$$M[2,3] = \left(\frac{k_2 \cos\phi_L^r}{k_2}\right) e^{+ik_2 \cos\phi_L^r z_1}$$

$$M[2,4] = -\left(\frac{k_2 \cos\phi_L^r}{k_2}\right) e^{-ik_2 \cos\phi_L^r z_1}$$

$$M[2,5] = \left(\frac{\gamma}{K_2}\right) e^{+iK_2 \cos\phi_S^r z_1}$$

$$M[2,6] = \left(\frac{\gamma}{K_2}\right) e^{-iK_2 \cos\phi_S^r z_1}$$

$$M[2,7] = M[2,8] = 0$$

$$M[3,1] = -\frac{i\mu_1}{k_1}\left(K_1^2 - 2\gamma^2\right)$$

$$M[3,2] = \frac{i\mu_1}{K_1}(2\gamma K_1 \cos\phi_S^r)$$

$$M[3,3] = \frac{i\mu_2}{k_2}\left(K_2^2 - 2\gamma^2\right) e^{+ik_2 \cos\phi_L^r z_1}$$

$$M[3,4] = \frac{i\mu_2}{k_2}\left(K_2^2 - 2\gamma^2\right) e^{-ik_2 \cos\phi_L^r z_1}$$

$$M[3,5] = \frac{i\mu_2}{K_2}(2\gamma K_2 \cos\phi_S^r) e^{+iK_2 \cos\phi_S^r z_1}$$

$$M[3,6] = -\frac{i\mu_2}{K_2}\left(2\gamma K_2 \cos\phi_S^r\right) e^{-iK_2 \cos\phi_S^r z_1}$$

$$M[3,7] = M[3,8] = 0$$

$$M[4,1] = \frac{i\mu_1}{k_1}\left(2\gamma k_1 \cos\phi_L^s\right)$$

$$M[4,2] = \frac{i\mu_1}{K_1}\left(K_1^2 - 2\gamma^2\right)$$

$$M[4,3] = \frac{i\mu_2}{k_2}\left(2\gamma k_2 \cos\phi_L^r\right) e^{+ik_2 \cos\phi_L^r z_1}$$

$$M[4,4] = -\frac{i\mu_2}{k_2}\left(2\gamma k_2 \cos\phi_L^r\right) e^{-ik_2 \cos\phi_L^r z_1}$$

$$M[4,5] = -\frac{i\mu_2}{K_2}\left(K_2^2 - 2\gamma^2\right) e^{+iK_2 \cos\phi_S^r z_1}$$

$$M[4,6] = -\frac{i\mu_2}{K_2}\left(K_2^2 - 2\gamma^2\right) e^{-iK_2 \cos\phi_S^r z_1}$$

$$M[4,7] = M[4,8] = M[5,1] = M[5,2] = 0$$

$$M[5,3] = \frac{i\mu_2}{k_2}\left(2\gamma k_2 \cos\phi_L^r\right) e^{+ik_2 \cos\phi_L^r z_2}$$

$$M[5,4] = -\frac{i\mu_2}{k_2}\left(2\gamma k_2 \cos\phi_L^r\right) e^{-ik_2 \cos\phi_L^r z_2}$$

$$M[5,5] = -\frac{i\mu_2}{K_2}\left(K_2^2 - 2\gamma^2\right) e^{+iK_2 \cos\phi_S^r z_2}$$

$$M[5,6] = -\frac{i\mu_2}{K_2}\left(K_2^2 - 2\gamma^2\right) e^{-iK_2 \cos\phi_S^r z_2}$$

$$M[5,7] = -\frac{i\mu_3}{k_3}\left(2\gamma k_3 \cos\phi_L^t\right)$$

$$
\begin{aligned}
M[5,8] &= \frac{i\mu_3}{K_3}\left(K_3^2 - 2\gamma^2\right) \\
M[6,1] &= M[6,2] = 0 \\
M[6,3] &= \frac{i\mu_2}{k_2}\left(K_2^2 - 2\gamma^2\right) e^{+ik_2\cos\phi_L^r z_2} \\
M[6,4] &= \frac{i\mu_2}{k_2}\left(K_2^2 - 2\gamma^2\right) e^{-ik_2\cos\phi_L^r z_2} \\
M[6,5] &= \frac{i\mu_2}{K_2}\left(2\gamma K_2\cos\phi_S^r\right) e^{+ik_2\cos\phi_L^r z_2} \\
M[6,6] &= -\frac{i\mu_2}{K_2}\left(2\gamma K_2\cos\phi_S^r\right) e^{-ik_2\cos\phi_L^r z_2} \\
M[6,7] &= -\frac{i\mu_3}{k_3}\left(K_3^2 - 2\gamma^2\right) \\
M[6,8] &= -\frac{i\mu_3}{K_3}\left(2\gamma K_3\cos\phi_S^t\right) \\
M[7,1] &= M[7,2] = 0 \\
M[7,3] &= \left(\frac{k_2\cos\phi_L^r}{k_2}\right) e^{ik_2\cos\phi_L^r z_2} \\
M[7,4] &= -\left(\frac{k_2\cos\phi_L^r}{k_2}\right) e^{-ik_2\cos\phi_L^r z_2} \\
M[7,5] &= \left(\frac{\gamma}{K_2}\right) e^{+iK_2\cos\phi_S^r z_2} \\
M[7,6] &= \left(\frac{\gamma}{K_2}\right) e^{-iK_2\cos\phi_S^r z_2} \\
M[7,7] &= -\left(\frac{k_3\cos\phi_L^r}{k_3}\right)
\end{aligned}
$$

$$
\begin{aligned}
M[7,8] &= -\left(\frac{\gamma}{K_3}\right) \\
M[8,1] &= M[8,2] = 0 \\
M[8,3] &= \left(\frac{\gamma}{k_2}\right) e^{+ik_2 \cos\phi_L^r z_2} \\
M[8,4] &= \left(\frac{\gamma}{k_2}\right) e^{-ik_2 \cos\phi_L^r z_2} \\
M[8,5] &= -\left(\frac{K_2 \cos\phi_S^r}{K_2}\right) e^{+iK_2 \cos\phi_S^r z_2} \\
M[8,6] &= \left(\frac{K_2 \cos\phi_S^r}{K_2}\right) e^{-iK_2 \cos\phi_S^r z_2} \\
M[8,7] &= -\left(\frac{\gamma}{k_3}\right) \\
M[8,8] &= \left(\frac{K_3 \cos\phi_S^t}{K_3}\right) \qquad (3.5)
\end{aligned}
$$

In the above matrix, we have already factored out the exponential terms in the transmitted and scattered fields. The exponential terms in the refracted field remain in the boundary conditions, complicating the expressions. The boundary condition matrix is extremely large and complex, but it is finite and can be manipulated to achieve a considerable amount of simplification. The first four rows are the boundary conditions applied to the interface between the layer and exterior medium, and the second four rows are the boundary conditions for the inner interface. Each column of the matrix represents components from a longitudinal or transverse displacement field, with the first two columns representing the scattered longitudinal and transverse displacement field contributions to the boundary conditions, in that order. By paying close attention to the row and column operations, we can reduce the above forty-eight, non-zero terms, to only eight non-zero or non-unity terms. It now only remains to express the incident planar displacement field in terms of the boundary conditions. The b column vector for the incident longitudinal displacement wave has components representing displacements and surface tractions at the two boundaries. There are no terms on the inner boundary since the external field is separated from the internal field by the interface layer. After factoring out the exponential function common to all terms, we arrive at the following expressions:

$$
\begin{aligned}
b[1] &= \left(\frac{\gamma}{k_1}\right) \\
b[2] &= \left(\frac{k_1 \cos\phi^i}{k_1}\right) \\
b[3] &= \frac{i\mu_1}{k_1}\left(K_1^2 - 2\gamma^2\right) \\
b[4] &= \frac{i\mu_1}{k_1}\left(2\gamma k_1 \cos\phi^i\right) \\
b[5] &= b[6] = b[7] = b[8] = 0 \qquad (3.6)
\end{aligned}
$$

The formal solution to the scattering of an incident longitudinal displacement field from a layered planar interface is now reduced to writing the coefficients of the potentials in terms of these stated boundary conditions. We will factor out of the matrix as many terms as possible and group with the modal coefficients, which we represented by the x column multiplying the matrix M and algebraically invert the system. The coefficients of the displacement potentials are then given by the algebraic ratio of the determinants of the boundary value matrix and the substitution matrices. We define the determinants as

$\Delta_i^{L,S,V}$, where $i = (0, 1, 2, .)$. The numbers in the expression for the determinant indicate the column of the matrix that b has been substituted into and the superscripts (L, S, V) indicate the type of incident displacement field. Therefore, $x_1 = \frac{\Delta_1^L}{\Delta_0}$ by Cramer's rule, represents the coefficient of the scattered longitudinal displacement potential due to an incident longitudinal displacement field. We are then left with solving for the coefficients in the potentials. The coefficients of the displacement potentials that we need to find are the scattered ones to solve the scattering or reflection problem for the planar interface. The coefficients of the scattered displacement fields can be written in terms of the determinants as:

$$
\begin{aligned}
a_1 &= \left(\frac{\Delta_1^L}{\Delta_0}\right) \\
b_1 &= \frac{K_1}{k_1} e^{iK_1 \cos\phi_S^s z_1} e^{ik_1 \cos\phi^i z_1} \left(\frac{\Delta_2^L}{\Delta_0}\right)
\end{aligned}
\tag{3.7}
$$

where the factor in front of the transverse potential is the ratio of terms factored out of the boundary condition matrix prior to taking the determinants. This has significant advantages over the numerical inversion of this system to solve for the coefficients. The expressions given can be manipulated algebraically to study the behavior as a function of the material parameters and geometry.

3.2.1 Scattered fields

When an incident longitudinal displacement wave is incident on a planar interface at an angle other than zero with respect to the direction of propagation, then scattered fields of longitudinal and transverse displacements will be generated. The scattered fields from the planar interface are written in terms of their displacement potentials, with the substitution made for the solutions for the coefficients giving:

$$\pi_L^s = \left(\frac{\Delta_1^L}{\Delta_0}\right) e^{-ik_1 \cos\phi_L^s z} e^{ik_1 \sin\phi_L^s x}$$

$$\pi_S^s = \left(\frac{K_1}{k_1} e^{iK_1 \cos\phi_S^s z_1} e^{ik_1 \cos\phi^i z_1} \frac{\Delta_2^L}{\Delta_0}\right) e^{-iK_1 \cos\phi_S^s z} e^{iK_1 \sin\phi_S^s x} \tag{3.8}$$

We can see from the displacement potentials that the coupling of these scattered displacement fields to the displacement fields inside of the interface layer and in the interior are completely contained within the determinant ratios. All other factors in the displacement potentials are dependent only on the external field variables, including the angle of propagation. If we take a look at the expression for Δ_0 (eqn. 3.9), as complicated as it appears, there is a clear structure to the algebraic solution that can be seen. First we should note that the subscripts on the components of the factors in the determinant actually represent the matrix element position after extensive simplification of the boundary condition matrix. Secondly, we note that through careful manipulation, we have not mixed the expressions for the interior field with those of the exterior field in the matrix manipulations. The first four terms in Δ_0 are coupling terms between the exterior field and interface layer, and the remaining four terms are coupling between the interface layer and the interior. This interface coupling makes logical sense, since the exterior and interior fields are separated by the interface layer, the dependence of interior and exterior fields must be through the properties of the interface layer. The expressions for $\Delta_0, \Delta_1^L, \Delta_2^L$ are given as follows, with the last four elements of the determinant identical for all three determi-

nants, as expected, since the relations between the interface layer and interior region must not be affected by substitution of the column vector *b* into the columns corresponding to components from the scattered field. The expressions for the determinants are:

$$\Delta_0 = -1 + D_{11}D_{78} + D_{22}D_{87} + D_{21}D_{77} + D_{12}D_{88} - (D_{11}D_{22} - D_{12}D_{21})(D_{77}D_{88} - D_{78}D_{87}) \tag{3.9}$$

$$D_{11} = \frac{-\tan\phi_{2s}}{(1+F_2^T)e^{iK_2 d\cos\phi_{2s}}}\left[1 - \frac{\left(\frac{\mu_2}{\mu_1}\right)\frac{(K_2d)^2}{2}\left(\left(\frac{\mu_2-\mu_1}{\mu_1}\right)(\sigma d)^2(1+\tan\phi_{1l}\tan\phi_{1s}) + \tan\phi_{1l}\tan\phi_{1s}\frac{(K_1d)^2}{2}\right)}{Den_1}\right]$$

$$D_{12} = \frac{1}{(1+F_2^T)e^{iK_2 d\cos\phi_{2s}}}\left[F_2^T + \frac{\left(\frac{\mu_2}{\mu_1}\right)\tan\phi_{1l}\tan\phi_{2s}\frac{(K_1d)^2}{2}\frac{(K_2d)^2}{2}}{Den_1}\right]$$

$$D_{21} = \frac{1}{(1+F_2^L)e^{ik_2 d\cos\phi_{2l}}}\left[F_2^L + \frac{\left(\frac{\mu_2}{\mu_1}\right)\tan\phi_{1s}\tan\phi_{2l}\frac{(K_1d)^2}{2}\frac{(K_2d)^2}{2}}{Den_1}\right]$$

$$D_{22} = \frac{\tan\phi_{2l}}{(1+F_2^L)e^{ik_2 d\cos\phi_{2l}}}\left[1 - \frac{\left(\frac{\mu_2}{\mu_1}\right)\frac{(K_2d)^2}{2}\left(\left(\frac{\mu_2-\mu_1}{\mu_1}\right)(\sigma d)^2(1+\tan\phi_{1l}\tan\phi_{1s}) + \tan\phi_{1l}\tan\phi_{1s}\frac{(K_1d)^2}{2}\right)}{Den_1}\right]$$

$$D_{77} = \frac{1}{2F_1^L}\left[F_2^L + \frac{\left(\frac{\mu_2}{\mu_3}\right)\tan\phi_{2l}\tan\phi_{3s}\frac{(K_2d)^2}{2}\frac{(K_3d)^2}{2}}{Den_2}\right]$$

$$D_{78} = \frac{-\tan\phi_{2l}}{2F_1^L}\left[1 - \frac{\left(\frac{\mu_2}{\mu_3}\right)\frac{(K_2d)^2}{2}\left(\left(\frac{\mu_2-\mu_3}{\mu_3}\right)(\sigma d)^2(1+\tan\phi_{3l}\tan\phi_{3s}) + \tan\phi_{3l}\tan\phi_{3s}\frac{(K_3d)^2}{2}\right)}{Den_2}\right]$$

$$D_{87} = \frac{\tan\phi_{2s}}{2F_1^T}\left[1 - \frac{\left(\frac{\mu_2}{\mu_3}\right)\frac{(K_2d)^2}{2}\left(\left(\frac{\mu_2-\mu_3}{\mu_3}\right)(\sigma d)^2(1+\tan\phi_{3l}\tan\phi_{3s}) + \tan\phi_{3l}\tan\phi_{3s}\frac{(K_3d)^2}{2}\right)}{Den_2}\right]$$

$$D_{88} = \frac{1}{2F_1^T}\left[F_2^T + \frac{\left(\frac{\mu_2}{\mu_3}\right)\tan\phi_{2s}\tan\phi_{3l}\frac{(K_2d)^2}{2}\frac{(K_3d)^2}{2}}{Den_2}\right]$$

$$Den_1 = \left[\left(\frac{\mu_2-\mu_1}{\mu_1}\right)^2 (\sigma d)^4 (1+\tan\phi_{1l}\tan\phi_{1s}) + \tan\phi_{1l}\tan\phi_{1s}\frac{(K_1 d)^2}{2}\left(\frac{(K_1 d)^2}{2} + 2\left(\frac{\mu_2-\mu_1}{\mu_1}\right)(\sigma d)^2\right)\right]$$

$$Den_2 = \left[\left(\frac{\mu_2-\mu_3}{\mu_3}\right)^2 (\sigma d)^4 (1+\tan\phi_{3l}\tan\phi_{3s}) + \tan\phi_{3l}\tan\phi_{3s}\frac{(K_3 d)^2}{2}\left(\frac{(K_3 d)^2}{2} + 2\left(\frac{\mu_2-\mu_3}{\mu_3}\right)(\sigma d)^2\right)\right]$$

$$F_2^T = \frac{e^{-iK_2 d\cos\phi_{2s}} + e^{iK_2 d\cos\phi_{2s}}}{e^{-iK_2 d\cos\phi_{2s}} - e^{iK_2 d\cos\phi_{2s}}} \qquad F_2^L = \frac{e^{-ik_2 d\cos\phi_{2l}} + e^{ik_2 d\cos\phi_{2l}}}{e^{-ik_2 d\cos\phi_{2l}} - e^{ik_2 d\cos\phi_{2l}}}$$

$$F_1^T = \frac{1}{e^{-iK_2 d\cos\phi_{2s}} - e^{iK_2 d\cos\phi_{2s}}} \qquad F_1^L = \frac{1}{e^{-ik_2 d\cos\phi_{2l}} - e^{ik_2 d\cos\phi_{2l}}}$$

$$\Delta_1^L = Mult_1^L\left(-1 + D_{11}D_{78} + D_{22}D_{87} + D_{21}D_{77} + D_{12}D_{88} - (D_{11}D_{22} - D_{12}D_{21})(D_{77}D_{88} - D_{78}D_{87})\right) \tag{3.10}$$

$$\begin{aligned}
D_{11} &= \frac{-\tan\phi_{2s}}{(1+F_2^T)e^{iK_2 d\cos\phi_{2s}}}\left[1 + \frac{\left(\frac{\mu_2}{\mu_1}\right)\frac{(K_2 d)^2}{2}\left(\left(\frac{\mu_2-\mu_1}{\mu_1}\right)(\sigma d)^2(1-\tan\phi_{1l}\tan\phi_{1s}) - \tan\phi_{1l}\tan\phi_{1s}\frac{(K_1 d)^2}{2}\right)}{Den_1}\right]\\
D_{12} &= \frac{1}{(1+F_2^T)e^{iK_2 d\cos\phi_{2s}}}\left[F_2^T - \frac{\left(\frac{\mu_2}{\mu_1}\right)\tan\phi_{1l}\tan\phi_{2s}\frac{(K_1 d)^2}{2}\frac{(K_2 d)^2}{2}}{Den_1}\right]\\
D_{21} &= \frac{1}{(1+F_2^L)e^{ik_2 d\cos\phi_{2l}}}\left[F_2^L + \frac{\left(\frac{\mu_2}{\mu_1}\right)\tan\phi_{1s}\tan\phi_{2l}\frac{(K_1 d)^2}{2}\frac{(K_2 d)^2}{2}}{Den_1}\right]\\
D_{22} &= \frac{\tan\phi_{2l}}{(1+F_2^L)e^{ik_2 d\cos\phi_{2l}}}\left[1 - \frac{\left(\frac{\mu_2}{\mu_1}\right)\frac{(K_2 d)^2}{2}\left(\left(\frac{\mu_2-\mu_1}{\mu_1}\right)(\sigma d)^2(1-\tan\phi_{1l}\tan\phi_{1s}) - \tan\phi_{1l}\tan\phi_{1s}\frac{(K_1 d)^2}{2}\frac{(K_2 d)^2}{2}\right)}{Den_1}\right]\\
Den_1 &= \left(\frac{\mu_2-\mu_1}{\mu_1}\right)^2(\sigma d)^4(1-\tan\phi_{1l}\tan\phi_{1s}) - \tan\phi_{1l}\tan\phi_{1s}\frac{(K_1 d)^2}{2}\left(\frac{(K_1 d)^2}{2} + 2\left(\frac{\mu_2-\mu_1}{\mu_1}\right)(\sigma d)^2\right)\\
Mult_1^L &= \frac{\left(\frac{\mu_2-\mu_1}{\mu_1}\right)^2(\sigma d)^4(1-\tan\phi_{1l}\tan\phi_{1s}) - \tan\phi_{1l}\tan\phi_{1s}\frac{(K_1 d)^2}{2}\left(\frac{(K_1 d)^2}{2} + 2\left(\frac{\mu_2-\mu_1}{\mu_1}\right)(\sigma d)^2\right)}{\left(\frac{\mu_2-\mu_1}{\mu_1}\right)^2(\sigma d)^4(1+\tan\phi_{1l}\tan\phi_{1s}) + \tan\phi_{1l}\tan\phi_{1s}\frac{(K_1 d)^2}{2}\left(\frac{(K_1 d)^2}{2} + 2\left(\frac{\mu_2-\mu_1}{\mu_1}\right)(\sigma d)^2\right)}
\end{aligned}$$

$$\Delta_2^L = Mult_2^L(-1 + D_{11}D_{78} + D_{22}D_{87} + D_{21}D_{77} + D_{12}D_{88} - (D_{11}D_{22} - D_{12}D_{21})(D_{77}D_{88} - D_{78}D_{87})) \tag{3.11}$$

$$\begin{aligned}
D_{11} &= \frac{-\tan\phi_{2s}\left(\left(\frac{\mu_2-\mu_1}{\mu_1}\right)(\sigma d)^2 + \frac{(K_1 d)^2}{2} - \left(\frac{\mu_2}{\mu_1}\right)\frac{(K_2 d)^2}{2}\right)}{(1+F_2^T)e^{iK_2 d\cos\phi_{2s}}\left(\left(\frac{\mu_2-\mu_1}{\mu_1}\right)(\sigma d)^2 + \frac{(K_1 d)^2}{2}\right)} \\
D_{12} &= \frac{F_2^T}{(1+F_2^T)e^{iK_2 d\cos\phi_{2s}}} \\
D_{21} &= \frac{F_2^L}{(1+F_2^L)e^{ik_2 d\cos\phi_{2l}}} \\
D_{22} &= \frac{\tan\phi_{2l}\left(\left(\frac{\mu_2-\mu_1}{\mu_1}\right)(\sigma d)^2 - \left(\frac{\mu_2}{\mu_1}\right)\frac{(K_2 d)^2}{2}\right)}{(1+F_2^L)e^{ik_2 d\cos\phi_{2l}}} \\
Mult_2^L &= \frac{-2(\sigma d)^2\tan\phi_{1l}\left(\frac{\mu_2-\mu_1}{\mu_1}\right)\left(\left(\frac{\mu_2-\mu_1}{\mu_1}\right)(\sigma d)^2 + \frac{(K_1 d)^2}{2}\right)}{\left(\frac{\mu_2-\mu_1}{\mu_1}\right)^2(\sigma d)^4(1+\tan\phi_{1l}\tan\phi_{1s}) + \tan\phi_{1l}\tan\phi_{1s}\frac{(K_1 d)^2}{2}\left(\frac{(K_1 d)^2}{2} + 2\left(\frac{\mu_2-\mu_1}{\mu_1}\right)(\sigma d)^2\right)}
\end{aligned}$$

3.2.2 Dynamic stress concentrations

The expressions for the scattered displacement potentials allow us to explicitly write the expressions for the dynamic stress concentrations. These can then be evaluated for varying elastic properties and geometry to determine the effect of an interface layer on the dynamic stresses. We normalize the dynamic stress concentrations by calculating the incident stress field and computing σ^*, the dynamic stress concentration [78]. There will be three terms contributing to σ^* and it is preferable to review the individual contributions in order to interpret the physics of the problem. Since the incident longitudinal displacement field will only produce shear stresses when the angle of incidence is nonzero, we will normalize the scattered shear stresses by the incident normal stress. The expressions for stress concentration are then given as follows:

$$
\begin{aligned}
\sigma^*_{zz} &= \frac{\left|\sigma^i_{zz} + \sigma^L_{zz} + \sigma^S_{zz}\right|}{\left|\sigma^i_{zz}\right|} \\
\sigma^L_{zz} &= \frac{i\mu_1}{k_1}\left(K_1^2 - 2\gamma^2\right)\left(\frac{\Delta^L_1}{\Delta_0}\right) e^{-ik_1 \cos\phi^i z} e^{ik_1 \sin\phi^i x} \\
\sigma^S_{zz} &= \frac{-i\mu_1}{K_1}\left(2\gamma K_1 \cos\phi^s_S\right) \\
&\quad \left(\frac{K_1}{k_1} e^{iK_1 \cos\phi^s_S z_1} e^{ik_1 \cos\phi^i z_1} \frac{\Delta^L_2}{\Delta_0}\right) e^{-iK_1 \cos\phi^s_S z} e^{iK_1 \sin\phi^s_S x}
\end{aligned} \tag{3.12}
$$

$$
\begin{aligned}
\sigma^*_{xz} &= \frac{\left|\sigma^L_{xz} + \sigma^S_{xz}\right|}{\left|\sigma^i_{zz}\right|} \\
\sigma^L_{xz} &= \frac{-i2\mu_1}{k_1}\left(\gamma k_1 \cos\phi^i\right)\left(\frac{\Delta^L_1}{\Delta_0}\right) e^{-ik_1 \cos\phi^s_L z} e^{ik_1 \sin\phi^s_L x}
\end{aligned}
$$

$$\sigma_{xz}^{S} = \frac{-i2\mu_1}{K_1}\left(\frac{K_1^2}{2}-\gamma^2\right)$$

$$\left(\frac{K_1}{k_1}e^{iK_1\cos\phi_S^s z_1}e^{ik_1\cos\phi^i z_1}\frac{\Delta_2^L}{\Delta_0}\right)e^{-iK_1\cos\phi_S^s z}e^{iK_1\sin\phi_S^s x} \tag{3.13}$$

Now that we have the solutions, we can discuss some applications at the end of the chapter. Again, we should emphasize that the effects of the interface and interior elastic media on the scattered stress fields are contained completely in the ratio of the determinants.

3.3 Incident coupled transverse wave

In the case of an incident transverse wave of the π_S type, we have for the incident field

$$\pi_S^i = e^{iK_1\cos\phi^i z}e^{iK_1\sin\phi^i x} \tag{3.14}$$

The same situation as with the incident longitudinal wave occurs and we need to solve the 8×8 system to calculate the coefficients of the potentials. The matrix M is unchanged, but the column vector b, which represents the incident field components of displacement and normal surface traction on the boundaries, is modified. The b column vector for the incident transverse wave has the following components:

$$b[1] = -\left(\frac{K_1\cos\phi^i}{K_1}\right)$$

$$b[2] = \left(\frac{\gamma}{K_1}\right)$$

$$b[3] = \frac{i\mu_1}{K_1}\left(2\gamma K_1\cos\phi^i\right)$$

$$b[4] = -\frac{i\mu_1}{K_1}\left(K_1^2-2\gamma^2\right)$$

$$b[5] = b[6] = b[7] = b[8] = 0 \tag{3.15}$$

We have already factored out the exponential terms in the incident, transmitted, and scattered fields. We then compute the determinant of the modified matrices and solve for the unknown coefficients A and B. The expression for Δ_0 remains unchanged from the previously given expression and the coefficients of the scattered fields can be written as:

$$A_1 = \frac{k_1}{K_1} e^{iK_1 \cos\phi^i z_1} e^{ik_1 \cos\phi_L^s z_1} \left(\frac{\Delta_1^S}{\Delta_0}\right)$$

$$B_1 = \left(\frac{\Delta_2^S}{\Delta_0}\right) \tag{3.16}$$

where the factor in front of the longitudinal potential is the ratio of terms factored out of the boundary condition matrix prior to taking the determinants.

3.3.1 Scattered fields

The scattered displacement fields due to the incident transverse displacement field will be both longitudinal and transverse as long as the angle of incidence is nonzero. The potentials are given, upon substitution of the expressions for the coefficients as:

$$\pi_L^s = \frac{k_1}{K_1} e^{iK_1 \cos\phi^i z_1} e^{ik_1 \cos\phi_L^s z_1} \left(\frac{\Delta_1^S}{\Delta_0}\right) e^{-ik_1 \cos\phi_L^s z} e^{ik_1 \sin\phi_L^s x}$$

$$\pi_S^s = \left(\frac{\Delta_2^S}{\Delta_0}\right) e^{-iK_1 \cos\phi^i z} e^{iK_1 \sin\phi^i x} \tag{3.17}$$

with the expressions for Δ_1^S and Δ_2^S given as:

$$\Delta_1^S = Mult_1^S\left(-1 + D_{11}D_{78} + D_{22}D_{87} + D_{21}D_{77} + D_{12}D_{88} - (D_{11}D_{22} - D_{12}D_{21})(D_{77}D_{88} - D_{78}D_{87})\right) \tag{3.18}$$

$$D_{11} = \frac{-\tan\phi_{2s}\left(\left(\frac{\mu_2-\mu_1}{\mu_1}\right)(\sigma d)^2 - \left(\frac{\mu_2}{\mu_1}\right)\frac{(K_2 d)^2}{2}\right)}{(1+F_2^T)\left(\frac{\mu_2-\mu_1}{\mu_1}\right)e^{iK_2 d\cos\phi_{2s}}(\sigma d)^2}$$

$$D_{12} = \frac{F_2^T}{(1+F_2^T)e^{iK_2 d\cos\phi_{2s}}}$$

$$D_{21} = \frac{F_2^L}{(1+F_2^L)e^{ik_2 d\cos\phi_{2l}}}$$

$$D_{22} = \frac{\tan\phi_{2l}\left(\left(\frac{\mu_2-\mu_1}{\mu_1}\right)(\sigma d)^2 + \frac{(K_1 d)^2}{2} - \left(\frac{\mu_2}{\mu_1}\right)\frac{(K_2 d)^2}{2}\right)}{(1+F_2^L)e^{ik_2 d\cos\phi_{2l}}\left(\left(\frac{\mu_2-\mu_1}{\mu_1}\right)(\sigma d)^2 + \frac{(K_1 d)^2}{2}\right)}$$

$$Mult_1^S = \frac{2(\sigma d)^2\tan\phi_{1l}\left(\left(\frac{\mu_2-\mu_1}{\mu_1}\right)^2(\sigma d)^2 + \left(\frac{\mu_2-\mu_1}{\mu_1}\right)\frac{(K_1 d)^2}{2}\right)}{\left(\frac{\mu_2-\mu_1}{\mu_1}\right)^2(\sigma d)^4(1+\tan\phi_{1l}\tan\phi_{1s}) + \tan\phi_{1l}\tan\phi_{1s}\frac{(K_1 d)^2}{2}\left(\frac{(K_1 d)^2}{2} + 2\left(\frac{\mu_2-\mu_1}{\mu_1}\right)(\sigma d)^2\right)}$$

$$\Delta_2^S = Mult_2^S\left(-1 + D_{11}D_{78} + D_{22}D_{87} + D_{21}D_{77} + D_{12}D_{88} - (D_{11}D_{22} - D_{12}D_{21})(D_{77}D_{88} - D_{78}D_{87})\right) \tag{3.19}$$

$$\begin{aligned}
D_{11} &= \frac{-\tan\phi_{2s}}{(1+F_2^T)e^{iK_2 d\cos\phi_{2s}}}\left[1 - \frac{\left(\frac{\mu_2}{\mu_1}\right)\frac{(K_2d)^2}{2}\left(\left(\frac{\mu_2-\mu_1}{\mu_1}\right)(\sigma d)^2(1-\tan\phi_{1l}\tan\phi_{1s}) - \tan\phi_{1l}\tan\phi_{1s}\frac{(K_1d)^2}{2}\right)}{Den_1}\right] \\
D_{12} &= \frac{1}{(1+F_2^T)e^{iK_2 d\cos\phi_{2s}}}\left[F_2^T + \frac{\left(\frac{\mu_2}{\mu_1}\right)\tan\phi_{1l}\tan\phi_{2s}\frac{(K_1d)^2}{2}\frac{(K_2d)^2}{2}}{Den_1}\right] \\
D_{21} &= \frac{1}{(1+F_2^L)e^{ik_2 d\cos\phi_{2l}}}\left[F_2^L - \frac{\left(\frac{\mu_2}{\mu_1}\right)\tan\phi_{1s}\tan\phi_{2l}\frac{(K_1d)^2}{2}\frac{(K_2d)^2}{2}}{Den_1}\right] \\
D_{22} &= \frac{\tan\phi_{2l}}{(1+F_2^L)e^{ik_2 d\cos\phi_{2l}}}\left[1 - \frac{\left(\frac{\mu_2}{\mu_1}\right)\frac{(K_2d)^2}{2}\left(\left(\frac{\mu_2-\mu_1}{\mu_1}\right)(\sigma d)^2(1-\tan\phi_{1l}\tan\phi_{1s}) - \tan\phi_{1l}\tan\phi_{1s}\frac{(K_1d)^2}{2}\right)}{Den_1}\right] \\
Den_1 &= \left(\frac{\mu_2-\mu_1}{\mu_1}\right)^2(\sigma d)^4(1-\tan\phi_{1l}\tan\phi_{1s}) - \tan\phi_{1l}\tan\phi_{1s}\frac{(K_1d)^2}{2}\left(\frac{(K_1d)^2}{2} + 2\left(\frac{\mu_2-\mu_1}{\mu_1}\right)(\sigma d)^2\right) \\
Mult_2^S &= \frac{\left(\frac{\mu_2-\mu_1}{\mu_1}\right)^2(\sigma d)^4(1-\tan\phi_{1l}\tan\phi_{1s}) - \tan\phi_{1l}\tan\phi_{1s}\frac{(K_1d)^2}{2}\left(\frac{(K_1d)^2}{2} + 2\left(\frac{\mu_2-\mu_1}{\mu_1}\right)(\sigma d)^2\right)}{\left(\frac{\mu_2-\mu_1}{\mu_1}\right)^2(\sigma d)^4(1+\tan\phi_{1l}\tan\phi_{1s}) + \tan\phi_{1l}\tan\phi_{1s}\frac{(K_1d)^2}{2}\left(\frac{(K_1d)^2}{2} + 2\left(\frac{\mu_2-\mu_1}{\mu_1}\right)(\sigma d)^2\right)}
\end{aligned}$$

3.3.2 Dynamic stress concentrations

We may now write the explicit expressions for the dynamic stress concentrations due to an incident transverse displacement field. These can then be evaluated to determine the effect of an interface layer. We normalize the dynamic stress concentrations by calculating the incident stress field and computing σ^*. In this case, all stresses will be normalized to the incident shear stress giving:

$$\begin{aligned}
\sigma_{zz}^* &= \frac{\left|\sigma_{zz}^L + \sigma_{zz}^S\right|}{\left|\sigma_{xz}^i\right|} \\
\sigma_{zz}^L &= \frac{i\mu_1}{k_1}\left(K_1^2 - 2\gamma^2\right)\frac{k_1}{K_1} e^{iK_1 \cos\phi^i z_1} e^{ik_1 \cos\phi_L^s z_1}\left(\frac{\Delta_1^S}{\Delta_0}\right) e^{-ik_1 \cos\phi_L^s z} e^{ik_1 \sin\phi_L^s x} \\
\sigma_{zz}^S &= \frac{-i\mu_1}{K_1}\left(2\gamma K_1 \cos\phi^i\right)\left(\frac{\Delta_2^S}{\Delta_0}\right) e^{-iK_1 \cos\phi_S^s z} e^{iK_1 \sin\phi^i x}
\end{aligned} \tag{3.20}$$

and

$$\begin{aligned}
\sigma_{xz}^* &= \frac{\left|\sigma_{xz}^i + \sigma_{xz}^L + \sigma_{xz}^S\right|}{\left|\sigma_{xz}^i\right|} \\
\sigma_{xz}^L &= \frac{-i}{k_1}\left(\gamma k_1 \cos\phi^i\right)\frac{k_1}{K_1} e^{iK_1 \cos\phi^i z_1} e^{ik_1 \cos\phi_L^s z_1}\left(\frac{\Delta_1^S}{\Delta_0}\right) e^{-ik_1 \cos\phi_L^s z} e^{ik_1 \sin\phi_L^s x} \\
\sigma_{xz}^S &= \frac{-i}{K_1}\left(\frac{K_1^2}{2} - \gamma^2\right)\left(\frac{\Delta_2^S}{\Delta_0}\right) e^{-iK_1 \cos\phi_S^s z} e^{iK_1 \sin\phi_S^s x}
\end{aligned} \tag{3.21}$$

3.4 Incident uncoupled transverse wave

If we have an incident transverse wave of unit amplitude incident on the planar layered interface, traveling primarily in the $\hat{z}$ direction and uncoupled to the longitudinal and first transverse field on the boundaries, only scattered, refracted, and transmitted displacement fields of the same transverse type will be generated. The incident π_V displacement field and scattered, refracted, and transmitted fields are given by the displacement potentials:

$$
\begin{aligned}
\pi_V^i &= e^{+iK_1 \cos\phi^i z} e^{iK_1 \sin\phi^i x} \\
\pi_V^s &= C_1 e^{-iK_1 \cos\phi_V^s z} e^{iK_1 \sin\phi_V^s x} \\
\pi_V^r &= C_2 e^{+iK_2 \cos\phi_V^r z} e^{iK_2 \sin\phi_V^r x} \qquad \tilde{\pi}_V^r = \tilde{C}_2 e^{-iK_1 \cos\phi_V^s z} e^{iK_1 \sin\phi_V^s x} \\
\pi_V^t &= C_3 e^{+iK_3 \cos\phi_V^t z} e^{iK_3 \sin\phi_V^t x}
\end{aligned}
\tag{3.22}
$$

By applying the boundary conditions, we arrive at the four equations and four unknowns to solve for. In matrix form we can write them as $Nx = c$. Filling the matrix N with the expressions for the displacements and normal surface tractions at the two interfaces gives the following values for the elements of the matrix N:

$$
\begin{aligned}
N[1,1] &= -1 \\
N[1,2] &= e^{+iK_2 \cos\phi_V^r z_1} \\
N[1,3] &= e^{-iK_2 \cos\phi_V^r z_1} \\
N[1,4] &= 0 \\
N[2,1] &= i\mu_1 (K_1 \cos\phi_V^s) \\
N[2,2] &= i\mu_2 (K_2 \cos\phi_V^r) e^{+iK_2 \cos\phi_V^r z_1}
\end{aligned}
$$

$$N[2,3] = -i\mu_2 \left(K_2 \cos \phi_V^r\right) e^{-iK_2 \cos\phi_V^r z_1}$$

$$N[2,4] = 0 \quad N[3,1] = 0$$

$$N[3,2] = i\mu_2 \left(K_2 \cos \phi_V^r\right) e^{+iK_2 \cos\phi_V^r z_2}$$

$$N[3,3] = -i\mu_2 \left(K_2 \cos \phi_V^r\right) e^{-iK_2 \cos\phi_V^r z_2}$$

$$N[3,4] = -i\mu_3 \left(K_3 \cos \phi_V^t\right)$$

$$N[4,1] = 0$$

$$N[4,2] = e^{+iK_2 \cos\phi_V^r z_2}$$

$$N[4,3] = e^{-iK_2 \cos\phi_V^r z_2}$$

$$N[4,4] = -1 \tag{3.23}$$

In the above matrix, we have already factored out the exponential terms in the transmitted and scattered fields. The exponential terms in the refracted field remains in the boundary conditions, complicating the expressions. It only remain to express the incident field in terms of the boundary conditions. The c column vector for the incident uncoupled transverse wave has the following components after factoring out the exponential function common to all terms:

$$c[1] = 1, \quad c[2] = i\mu_1 \left(K_1 \cos \phi^i\right), \quad c[3] = c[4] = 0 \tag{3.24}$$

and the coefficients of the potentials can be written in terms of the determinants as:

$$C_1 = \left(\frac{\Delta_1^V}{\Delta_0}\right)$$

$$C_2 = e^{-iK_2 \cos\phi_V^r z_2} e^{+iK_1 \cos\phi^i z_1} \left(\frac{\Delta_2^V}{\Delta_0}\right)$$

$$\tilde{C}_2 = e^{iK_2 \cos\phi_V^r z_2} e^{+iK_1 \cos\phi^i z_1} \left(\frac{\tilde{\Delta}_2^V}{\Delta_0}\right)$$

$$C_3 = e^{-iK_3 \cos\phi_V^t z_2} e^{+iK_1 \cos\phi^i z_1} \left(\frac{\Delta_3^V}{\Delta_0}\right) \tag{3.25}$$

3.4.1 Scattered field and stress concentration

The scattered field from the planar interface will be only transverse. The potential is given below, after substitution of the expression for the coefficient as

$$\pi_V^s = \left(\frac{\Delta_1^V}{\Delta_0}\right) e^{-iK_1 \cos\phi^i z} e^{iK_1 \sin\phi^i x} \tag{3.26}$$

with the expressions for Δ_0, Δ_1^V given as follows:

$$\Delta_1^V = F_1 e^{iK_2 d \cos\phi_V^r} - F_2 e^{-iK_2 d \cos\phi_V^r}$$

$$\Delta_0 = F_1 F_2 e^{-iK_2 d \cos\phi_V^r} - e^{iK_2 d \cos\phi_V^r}$$

$$F_1 = \frac{\left(K_2 \cos\phi_V^r - \frac{\mu_1}{\mu_2} K_1 \cos\phi^i\right)}{\left(K_2 \cos\phi_V^r + \frac{\mu_1}{\mu_2} K_1 \cos\phi^i\right)}$$

$$F_2 = \frac{\left(K_2 \cos\phi_V^r - \frac{\mu_3}{\mu_2} K_3 \cos\phi_V^t\right)}{\left(K_2 \cos\phi_V^r + \frac{\mu_3}{\mu_2} K_3 \cos\phi_V^t\right)} \tag{3.27}$$

The expression for the scattered field can now be used to explicitly write the dynamic stress concentration as:

$$\begin{aligned} \sigma_{yz}^{*} &= \frac{\left|\sigma_{yz}^{i} + \sigma_{yz}^{V}\right|}{\left|\sigma_{yz}^{i}\right|} \\ \sigma_{yz}^{V} &= i\mu_1 \left(-K_1 \cos\phi^i\right) \left(\frac{\Delta_1^V}{\Delta_0}\right) e^{-iK_1 \cos\phi^i z} e^{iK_1 \sin\phi^i x} \end{aligned} \quad (3.28)$$

3.5 Applications to layered composites

The intended applications for this analysis of the dynamic stress concentrations, due to an interface layer between two elastic half spaces, are to laminated composite materials and as a comparison to the other two geometries under study. For an incident longitudinal displacement field, the variation of dynamic stress concentration with $k_1 d$, where k_1 is the frequency divided by the longitudinal propagation velocity in the exterior field and d is the thickness of the interface layer, can be easily achieved through simply varying the values of the field variables.

Two common layered structures that demonstrate some well understood behavior of the dynamic stress concentrations follow. As in figure 3.1, we will choose three materials for the exterior, interface, and interior. The first configuration is a polymer, cement, and steel laminate in that order for exterior, interface, and interior. The second configuration is a steel, aluminum, and steel laminate. In figure 3.2, we show the total normal and shear dynamic stress concentrations for the first case, and in figures 3.3 and 3.4 the individual contributions from the longitudinal and transverse scattered fields are shown. In figure 3.5, the normal and shear dynamic stress concentrations are shown for the second case, and in figures 3.6 and 3.7 the contributions due to the longitudinal and transverse scattered fields are given.

In the first case, we have an interface layer of cement between an exterior polymer-like medium and an interior steel medium. A longitudinal (compressional) displacement wave of unit amplitude is incident at an angle ϕ and the dynamic stresses are plotted as a function of incident angle ϕ and $k_1 d$. The incident field has only a normal stress component at $\phi = 0$ and all terms are normalized to normal stress with the exception of the total normal dynamic stress concentration which includes the incident stress. Figure 3.2 shows that for small values $k_1 d$ the normal stress is equal and of opposite sign to the incident normal stress giving a magnitude of two. At small angles of incidence, figure 3.2 shows that the shear stresses go to zero and the longitudinal and transverse fields are uncoupled as expected. The remaining figures 3.3-3.4 show the longitudinal and transverse contributions to the normal and shear stresses. In all cases, the shear stresses go to zero for small angles and the coupling between longitudinal and transverse fields show a strong angular dependence. The use of steel, which is very rigid and dense in comparison to the exterior polymer, creates a reflecting surface for small values of $k_1 d$. We can think of the steel as perfectly rigid in comparison to the polymer and a displacement incident upon that surface would be perfectly reflected.

The second case also resembles a special one for small values of $k_1 d$. For a steel-aluminum-steel laminate, the small values of $k_1 d$ indicate a very thin sheet of aluminum separating the steel exterior and interior. As we see from figure 3.5, the total normal stresses go to unity and the shear stresses to zero as if the interface were not present. Figures 3.6-3.7 display the contributions to the normal and shear stresses from the longi-

tudinal and transverse displacement fields, showing the same behavior at the extremes of $k_1 d$ and ϕ.

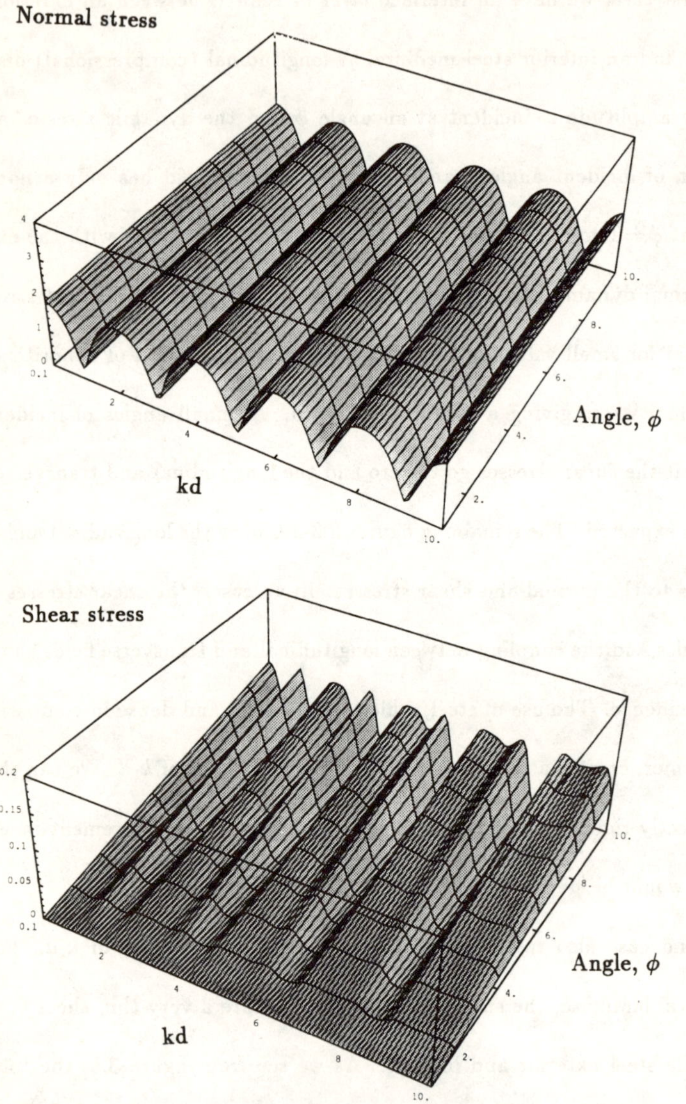

Figure 3.2: Total dynamic stress concentrations: polymer-cement-steel.

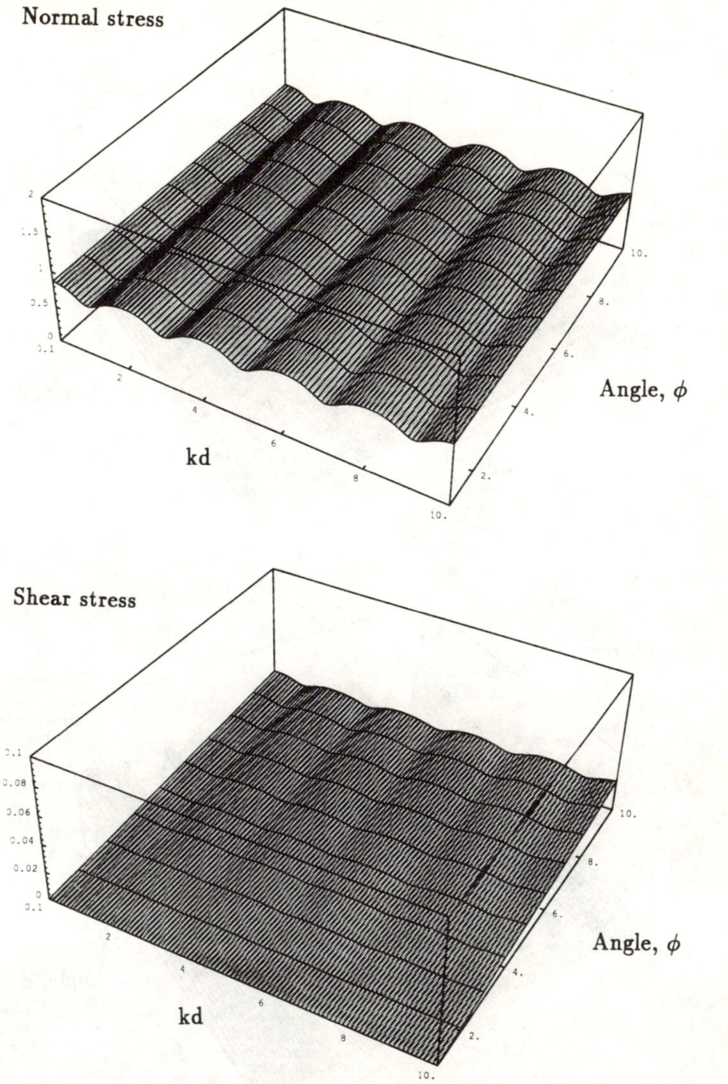

Figure 3.3: Longitudinal dynamic stress concentrations: polymer-cement-steel.

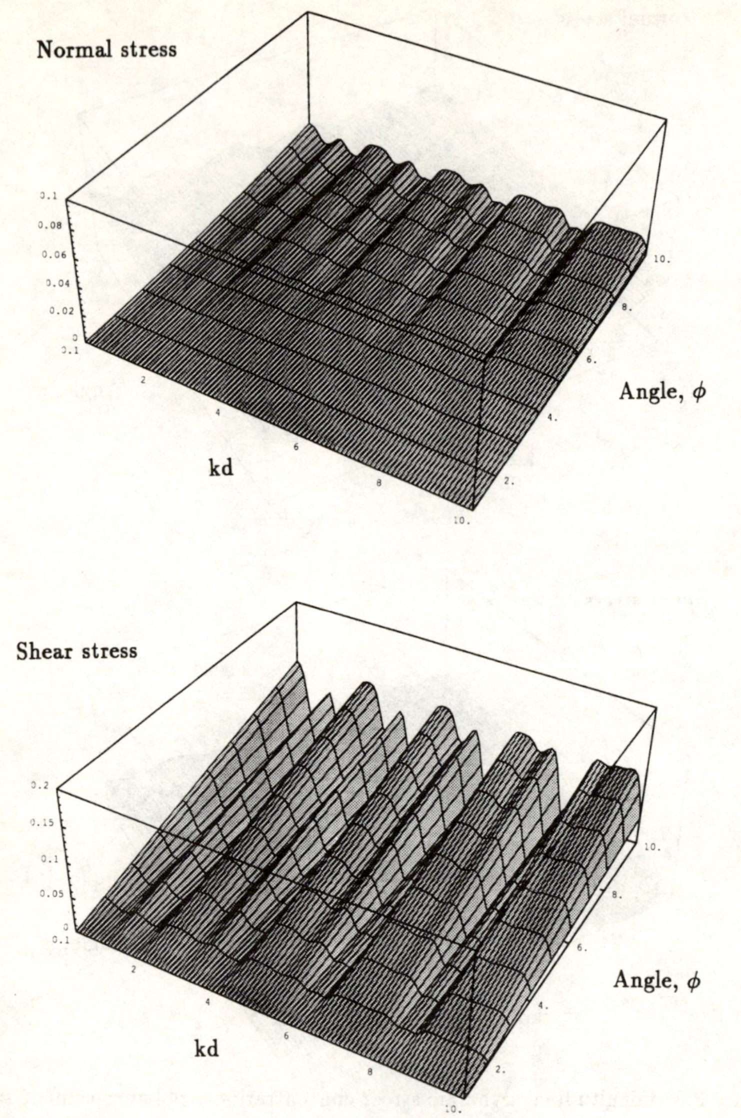

Figure 3.4: Transverse dynamic stress concentrations: polymer-cement-steel.

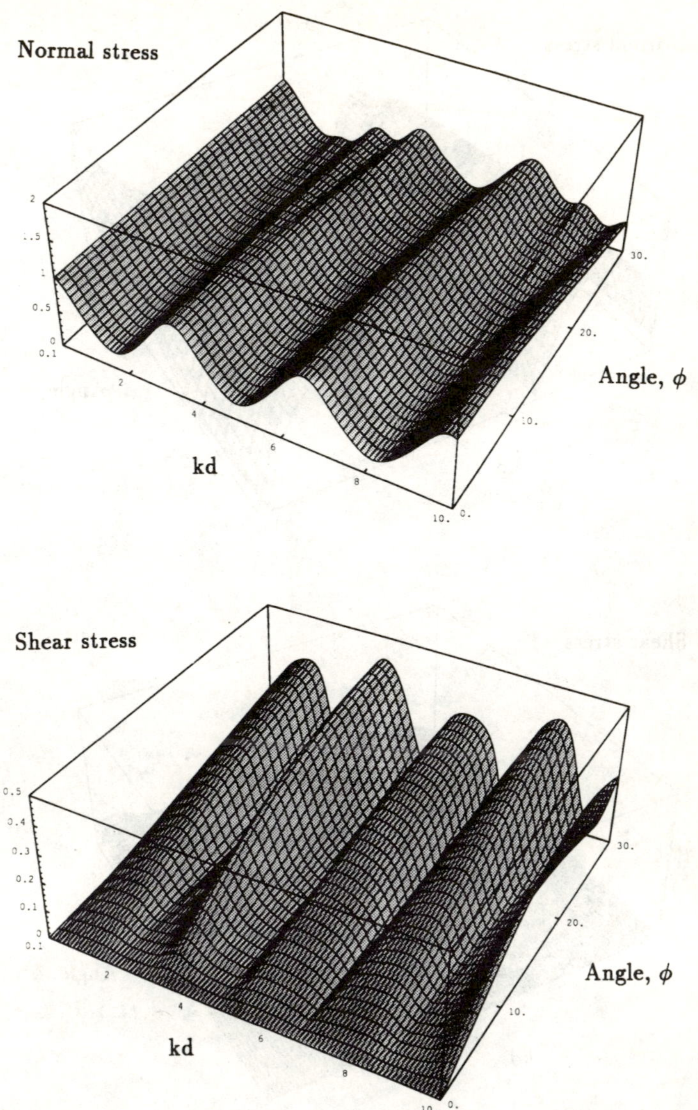

Figure 3.5: Total dynamic stress concentrations: steel-aluminum-steel.

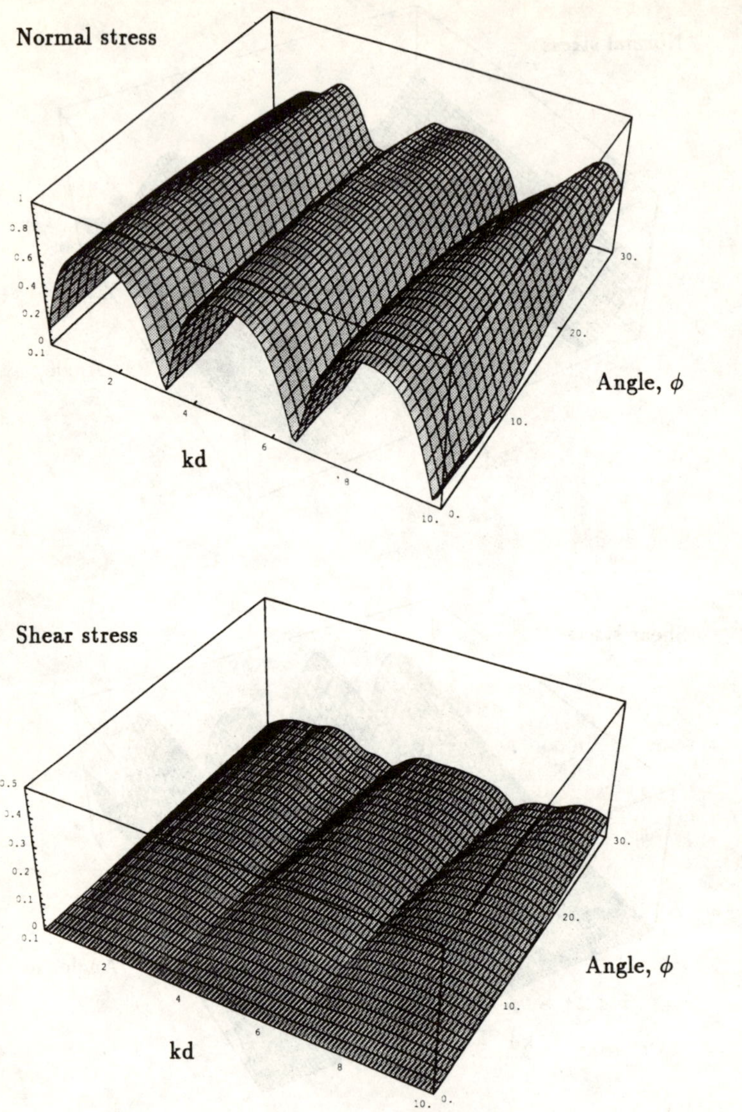

Figure 3.6: Longitudinal dynamic stress concentrations: steel-aluminum-steel.

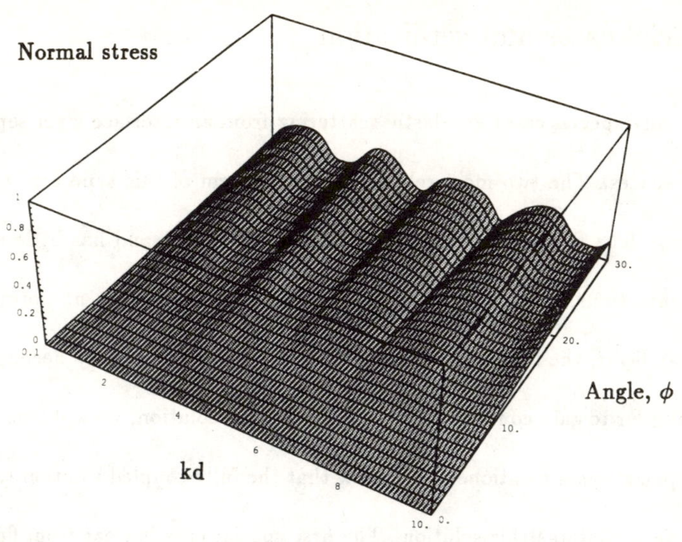

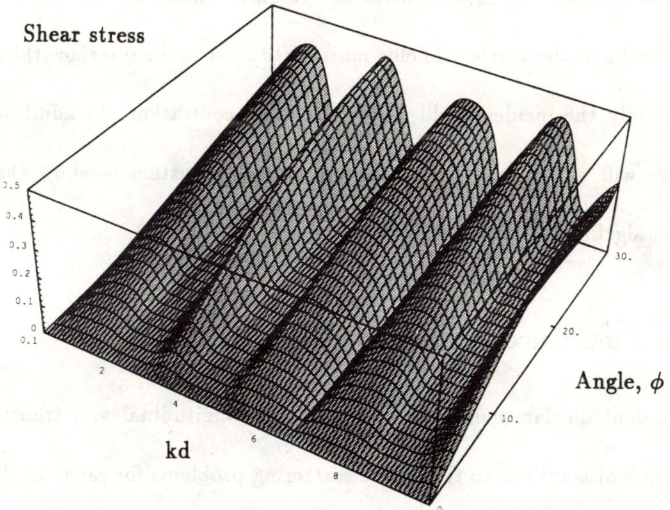

Figure 3.7: Transverse dynamic stress concentrations: steel-aluminum-steel.

3.6 Special cases and verification

There are many special cases for elastic scattering from an interface layer separating two elastic half spaces. The two most related to the problem of this type are: at low values of kd, the interface is small in comparison to the wavelength and has little effect on the stress concentrations, and when the incident displacement field is at normal incidence to the planar layer, the longitudinal and transverse fields completely decouple from one another. In order to gain confidence in the fully coupled solution, we will look at the latter of the two special cases mentioned, and show that the fully coupled solution to the layered planar interface contains this solution. The first special case is clear from figure 3.5, the limiting values of dynamic stress concentration are unity, since the external and internal media are identical and the scattered fields must go to zero as the interface thickness goes to zero leaving only the incident field in the stress concentration. In addition to these verifications, we will numerically invert the fully coupled system to show the identical results with the algebraically inverted system.

3.6.1 Normal incidence to the planar interface

A normally incident displacement field decouples the longitudinal and transverse fields allowing independent solutions to the elastic scattering problems for each. In this section, we compare the normal incidence solution for an incident longitudinal displacement field with the fully coupled non-normal incidence solution previously presented. Using the second case discussed in the previous section, a steel-aluminum-steel laminate, we can calculate the normal dynamic stress concentration and plot as a function of k_1d. The dynamic stress concentration, shown in figure 3.8, is the solution to the four equations and four unknowns for the planar interface layer with only longitudinal fields present, and is identical to the full solution given in figure 3.5 for the $\phi = 0$ plane.

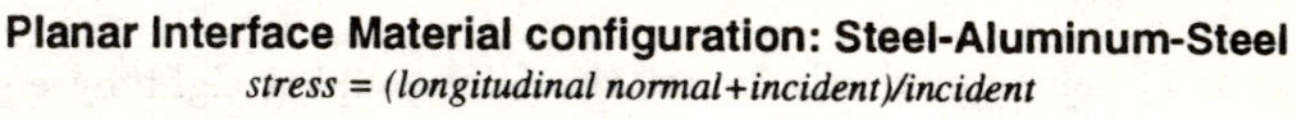

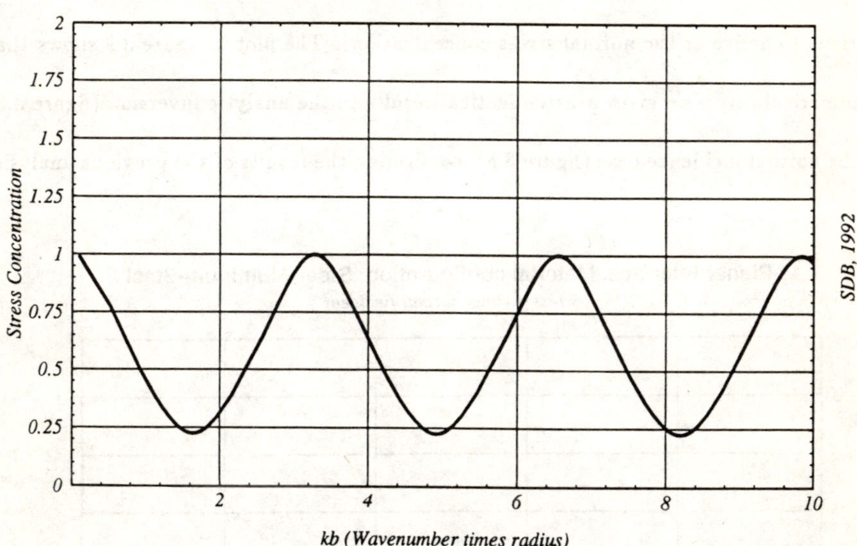

Figure 3.8: Normal incidence longitudinal dynamic stress concentration.

3.6.2 Numerical inversion

As a check with the results of the previous section, we take the second case of a steel-aluminum-steel laminate with an angle of incidence of the longitudinal displacement field of 0.1 degrees, and then numerically invert the fully coupled eight by eight system of equations to arrive at the normal stress concentrations. The plot in figure 3.9 shows that the numerical inversion gives nearly identical results to the analytic inversion (figure 3.5) and the normal incidence case (figure 3.8), confirming the results of the previous analysis.

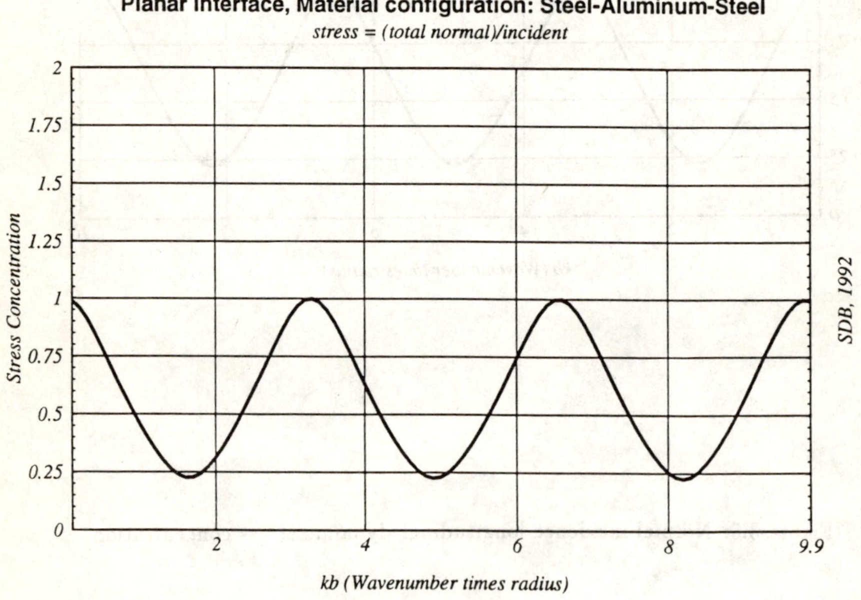

Figure 3.9: Numerical matrix inversion for dynamic stress concentrations.

Chapter 4

LAYERED CYLINDRICAL INCLUSION

4.1 Introduction

In the analysis of the scattering of elastic waves from a cylindrically layered interface embedded in an elastic medium, we are interested in modeling composites and the interface structures that develop as a result of processing or are intentionally introduced between the fiber and matrix of the composite. The geometry and field definitions for this case are given in figure 4.1 for reference. One example of this type of structure is in the chemical vapor deposition of boron or silicon-carbide fibers. The core fiber, usually carbon, serves as a substrate for deposition of the shell material. It is the shell of boron or silicon-carbide that is the actual load bearing part of the composite fiber. Another example of this structure is when tungsten wire is placed in a steel matrix. An interface zone between the fiber and matrix occurs, giving rise to a distinct shell coating the tungsten fibers. There are numerous other examples, most having to do with fiber reinforcement of materials. Our focus, in the layered cylindrical interface, will be on the study of fiber reinforced composites.

In this section, we will review the formal solution to elastic wave scattering from a layered elastic cylinder imbedded in an elastic medium, providing the analytic expressions for the scattered displacement fields. Three different incident displacement fields will be considered as incident on the interface layer originating in the exterior field: a longitudinal displacement field, a coupled transverse displacement field, and an uncoupled transverse

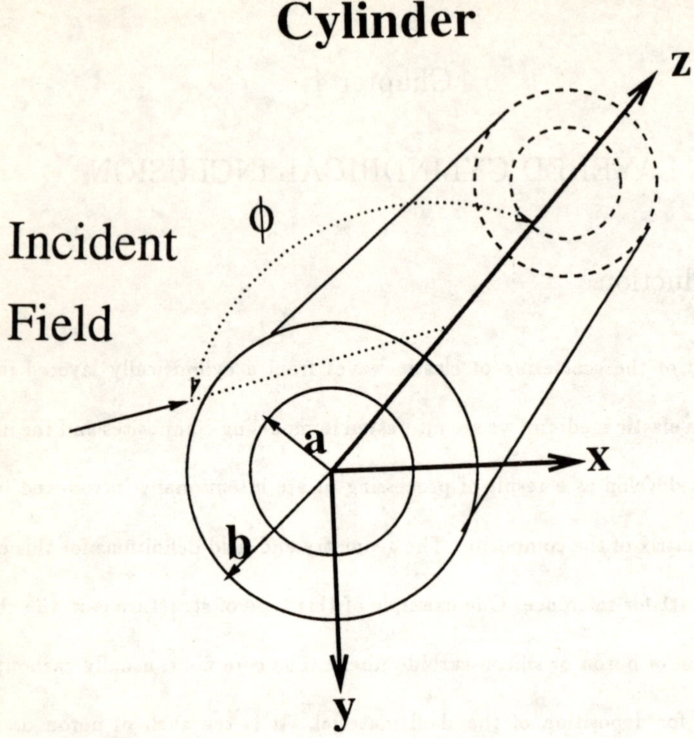

Figure 4.1: Problem geometry for the layered cylindrical interface.

displacement field. In this case, we have taken $\vec{a} = \hat{z}$ in computing the curls of the transverse displacement potentials. As discussed in appendix B, no other choice is possible for $\vec{a}$ in cylindrical coordinates.

To solve the scattering problem, we will need to set up the boundary conditions formulated in chapter 2 for the cylindrical boundary. The displacements and normal surface tractions are the appropriate field variables that must be continuous across these cylindrical boundaries. The boundary conditions involve derivatives of cylindrical Bessel and Hankel functions, and in order to simplify the expressions, we redefine these derivative by forming a log-like ratio of the derivative times argument over the function. For example, the cylindrical Bessel function ratio $k_1 b \frac{J_n'(k_1 b)}{J_n(k_1 b)}$ is rewritten as J_{b1}^L. Below we will use this notation, but not substitute in the appropriate radial functions quite yet. An intermediate symbolic step is necessary to write the expression for the displacements and normal surface tractions prior to evaluation at the boundaries. Rather than J we will use the symbol $Z_n^{L,S,V}$ to represent the log-like derivative of the radial functions, identifying the type of field by a superscript L, S, V.

While this may seem a tedious first step to take, the writing of the displacements and normal surface tractions in this form, then evaluating at the appropriate boundaries, eliminates a possible source of error in the manipulation of these lengthy expressions. To restate the definitions of some of the terms in the expressions, the subscript n refers to the summation index of the series solution to the Helmholtz equation. The displacement and stress fields will be an infinite series with a sum over the index n. The longitudinal displacement potential is written as π_L, coupled transverse as π_S, and uncoupled transverse as π_V. In addition, μ is the rigidity and the normal surface tractions are represented with the standard stress symbols $\sigma_{rr}, \sigma_{r\theta}, \sigma_{rz}$. The general expressions are then given as:

$$\sigma_{rr} = 2\mu\left[\left(\frac{i}{kr^2}\right)\left(\frac{(Kr)^2}{2} - n^2 + Z_n^L - (\gamma r)^2\right)\pi_L + \left(\frac{i}{Kr^2}\right)(in)\left(1 - Z_n^S\right)\pi_S - \left(\frac{i}{Kr}\right)^2 (i\gamma)\left(n^2 - Z_n^V - (Kr\sin\phi_v)^2\right)\pi_V\right] \tag{4.1}$$

$$\sigma_{r\theta} = 2\mu\left[\left(\frac{i}{kr^2}\right)(in)\left(1 - Z_n^L\right)\pi_L + \left(\frac{i}{Kr^2}\right)\left(n^2 - Z_n^S - \frac{(Kr\sin\phi_s)^2}{2}\right)\pi_S - \left(\frac{i}{Kr}\right)^2 (in)(i\gamma)\left(Z_n^V - 1\right)\pi_V\right]$$

$$\sigma_{rz} = 2\mu\left[\left(\frac{-i}{kr^2}\right)(i\gamma r)\left(Z_n^L\right)\pi_L + \left(\frac{-i}{Kr^2}\right)(in)\left(\frac{i\gamma r}{2}\right)\pi_S - \left(\frac{i}{Kr}\right)^2\left(\frac{(Kr)^2}{r}\right)\left(Z_n^V\right)\left(\frac{1}{2} - \cos^2\phi_v\right)\pi_V\right]$$

$$\begin{aligned}
u_r &= \left(\frac{-i}{kr}\right) Z_n^L \pi_L + \left(\frac{-i}{Kr}\right)(in)\,\pi_S - \left(\frac{i}{Kr}\right)^2 (i\gamma r)\, Z_n^V \pi_V \\
u_\theta &= \left(\frac{-i}{kr}\right)(in)\,\pi_L + \left(\frac{-i}{Kr}\right)\left(-Z_n^S\right)\pi_S - \left(\frac{i}{Kr}\right)^2 (i\gamma r)(in)\,\pi_V \\
u_z &= \left(\frac{-i}{kr}\right)(i\gamma r)\,\pi_L + \quad 0 \quad - \left(\frac{i}{Kr}\right)^2 (Kr\sin\phi_v)^2\,\pi_V
\end{aligned} \tag{4.2}$$

$$\begin{aligned}
Z_n^L &= \left(\frac{(kr\sin\phi_l)J_n'(kr\sin\phi_l)}{J_n(kr\sin\phi_l)}\right) \quad or \quad \left(\frac{(kr\sin\phi_l)H_n^{(1,2)\prime}(kr\sin\phi_l)}{H_n^{(1,2)}(kr\sin\phi_l)}\right) \\
Z_n^S &= \left(\frac{(Kr\sin\phi_s J_n'(Kr\sin\phi_s)}{J_n(Kr\sin\phi_s)}\right) \quad or \quad \left(\frac{(Kr\sin\phi_s H_n^{(1,2)\prime}(Kr\sin\phi_s)}{H_n^{(1,2)}(Kr\sin\phi_s)}\right) \\
Z_n^V &= \left(\frac{(Kr\sin\phi_v)J_n'(Kr\sin\phi_s)}{J_n(Kr\sin\phi_s)}\right) \quad or \quad \left(\frac{(Kr\sin\phi_v)H_n^{(1,2)\prime}(Kr\sin\phi_s)}{H_n^{(1,2)}(Kr\sin\phi_s)}\right)
\end{aligned} \tag{4.3}$$

The bounding surfaces are constant radial surfaces located at $r = a$ for the inner boundary and $r = b$ for the outer boundary, extending over the entire $\hat{z}$ axis. Incident longitudinal and transverse displacement fields are taken as propagating in the positive $\hat{z}$ and $\hat{x}$ directions. The value of γ is, using Snell's law as in the planar interface, determined by the direction of propagation and can be thought of as the projection of the wavenumber times propagation vector on the $\hat{z}$ axis, perpendicular to the primary direction of propagation and constant for all fields.

The proper boundary conditions, as discussed in chapter 2, are the three components of displacement and the radial components of stress given above. This gives at most six boundary conditions for each of the cylindrical boundaries. We assume that the angle of incidence is small enough that we can treat the π_V displacement wave as decoupled from the π_L and π_S displacement waves. This reduces the problem to a system of four equations and a system of two equations for each cylindrical boundary. The two boundary or two interface problem will therefore lead to a system of eight equations and eight unknowns for the π_L and π_S displacement potentials, and a system of four equations and four unknowns for the π_V displacement potentials.

We can write these sets of boundary conditions in symbolic form as $Mx = b$ for the incident longitudinal and coupled transverse field, and $Nx = c$ for the incident second transverse field. The algebraic solutions will then be written in terms of the ratio of the determinants via Cramer's rule.

4.2 Incident longitudinal wave

If we have an incident longitudinal (compressional-ultrasonic) wave of unit amplitude traveling in the positive $\hat{z}$ and $\hat{x}$ direction, scattered, refracted, and transmitted displacement fields will be generated (see figure 4.1). We begin by writing the incident planar displacement wave as the displacement potential

$$\pi_L^i = \sum_{n=0}^{\infty} \epsilon_n i^n J_n(k_1 r \sin\phi^i) e^{in\theta} e^{ik_1 z \cos\phi^i} \tag{4.4}$$

The eight unknown displacement potentials, defined in chapter 2, are given below after substitution for the propagation direction of the incident displacement field and assumed direction of propagation for the other fields. The angles that are defined in the potentials are the acute angles to the $\hat{z}$ axis as in the planar problem. The scattered longitudinal field propagates at the same angle as the incident longitudinal field by Snell's law, therefore the angles ϕ_L^s and ϕ^i will be equal. The displacement potentials can then be written as:

$$\pi_L^s = \sum_{n=0}^{\infty} a_{1n} i^n H_n^{(1)}(k_1 r \sin\phi_L^s) e^{in\theta} e^{ik_1 z \cos\phi_L^s}$$

$$\pi_S^s = \sum_{n=0}^{\infty} b_{1n} i^n H_n^{(1)}(K_1 r \sin\phi_S^s) e^{in\theta} e^{iK_1 z \cos\phi_S^s}$$

$$\pi_L^r = \sum_{n=0}^{\infty} a_{2n} i^n H_n^{(1)}(k_1 r \sin\phi_L^r) e^{in\theta} e^{ik_1 z \cos\phi_L^r}$$

$$\pi_S^r = \sum_{n=0}^{\infty} b_{2n} i^n H_n^{(1)}(K_2 r \sin\phi_S^r) e^{in\theta} e^{iK_2 z \cos\phi_S^r}$$

$$\tilde{\pi}_L^r = \sum_{n=0}^{\infty} \tilde{a}_{2n} i^n H_n^{(2)}(k_2 r \sin\phi_L^r) e^{in\theta} e^{ik_2 z \cos\phi_L^r}$$

$$\tilde{\pi}_S^r = \sum_{n=0}^{\infty} \tilde{b}_{2n} i^n H_n^{(2)}(K_2 r \sin\phi_S^r) e^{in\theta} e^{iK_2 z \cos\phi_S^r}$$

$$\pi_L^t = \sum_{n=0}^{\infty} a_{3n} i^n J_n(k_3 r \sin\phi_L^t) e^{in\theta} e^{ik_{3z}\cos\phi_L^t}$$

$$\pi_S^t = \sum_{n=0}^{\infty} b_{3n} i^n J_n(K_3 r \sin\phi_S^t) e^{in\theta} e^{iK_{3z}\cos\phi_S^t} \tag{4.5}$$

By applying the boundary conditions at the two cylindrical interfaces, we arrive at the eight equations for displacement and normal surface tractions and eight unknown modal coefficients of the displacement potentials to solve for. The radial functions in the interior and exterior regions are evaluated at the boundaries and factored out of the boundary conditions, and the radial functions for the shell region evaluated at the inner boundary are factored out as well. The matrix M has the following set of terms for bounding surfaces at $r = a$ and $r = b$:

$$M[1,1] = -in$$

$$M[1,2] = (K_1 b \sin\phi_S^s)\frac{H_n^{(1)\prime}(K_1 b \sin\phi_S^s)}{H_n^{(1)}(K_1 b \sin\phi_S^s)}$$

$$M[1,3] = in\frac{H_n^{(1)}(k_2 b \sin\phi_L^r)}{H_n^{(1)}(k_2 a \sin\phi_L^r)}$$

$$M[1,4] = in\frac{H_n^{(2)}(k_2 b \sin\phi_L^r)}{H_n^{(2)}(k_2 a \sin\phi_L^r)}$$

$$M[1,5] = -(K_2 b \sin\phi_S^r)\frac{H_n^{(1)\prime}(K_2 b \sin\phi_S^r)}{H_n^{(1)}(K_2 a \sin\phi_S^r)}$$

$$M[1,6] = -(K_2 b \sin\phi_S^r)\frac{H_n^{(2)\prime}(K_2 b \sin\phi_S^r)}{H_n^{(2)}(K_2 a \sin\phi_S^r)}$$

$$M[1,7] = M[1,8] = 0$$

$$M[2,1] = -(k_1 b \sin\phi_L^s)\frac{H_n^{(1)\prime}(k_1 b \sin\phi_L^s)}{H_n^{(1)}(k_1 b \sin\phi_L^s)}$$

$$M[2,2] = -in$$

$$M[2,3] = (k_2 b \sin\phi_L^r)\frac{H_n^{(1)\prime}(k_2 b \sin\phi_L^r)}{H_n^{(1)}(k_2 a \sin\phi_L^r)}$$

$$M[2,4] = (k_2 b \sin\phi_L^r)\frac{H_n^{(2)\prime}(k_2 b \sin\phi_L^r)}{H_n^{(2)}(k_2 a \sin\phi_L^r)}$$

$$M[2,5] = in\frac{H_n^{(1)}(K_2 b \sin\phi_S^r)}{H_n^{(1)}(K_2 a \sin\phi_S^r)}$$

$$M[2,6] = in\frac{H_n^{(2)}(K_2 b \sin\phi_S^r)}{H_n^{(2)}(K_2 a \sin\phi_S^r)}$$

$$M[2,7] = M[2,8] = 0$$

$$M[3,1] = in\mu_1\left(1-(k_1 b \sin\phi_L^s)\frac{H_n^{(1)\prime}(k_1 b \sin\phi_L^s)}{H_n^{(1)}(k_1 b \sin\phi_L^s)}\right)$$

$$M[3,2] = -\mu_1\left((K_1 b \sin\phi_S^s)\frac{H_n^{(1)\prime}(K_1 b \sin\phi_S^s)}{H_n^{(1)}(K_1 b \sin\phi_S^s)} - n^2 + \frac{(K_1 b)^2}{2}\sin^2\phi_S^s\right)$$

$$M[3,3] = -in\mu_2\left(1-(k_2 b \sin\phi_L^r)\frac{H_n^{(1)\prime}(k_2 b \sin\phi_L^r)}{H_n^{(1)}(k_2 b \sin\phi_L^r)}\right)\frac{H_n^{(1)}(k_2 b \sin\phi_L^r)}{H_n^{(1)}(k_2 a \sin\phi_L^r)}$$

$$M[3,4] = -in\mu_2\left(1-(k_2 b \sin\phi_L^r)\frac{H_n^{(2)\prime}(k_2 b \sin\phi_L^r)}{H_n^{(2)}(k_2 b \sin\phi_L^r)}\right)\frac{H_n^{(2)}(k_2 b \sin\phi_L^r)}{H_n^{(2)}(k_2 a \sin\phi_L^r)}$$

$$M[3,5] = \mu_2\left((K_2 b \sin\phi_S^r)\frac{H_n^{(1)\prime}(K_2 b \sin\phi_S^r)}{H_n^{(1)}(K_2 b \sin\phi_S^r)} - n^2 + \frac{(K_2 b)^2}{2}\sin^2\phi_S^r\right)\frac{H_n^{(1)}(K_2 b \sin\phi_S^r)}{H_n^{(1)}(K_2 a \sin\phi_S^r)}$$

$$M[3,6] = \mu_2\left((K_2 b\sin\phi_S^r)\frac{H_n^{(2)\prime}(K_2 b\sin\phi_S^r)}{H_n^{(2)}(K_2 b\sin\phi_S^r)} - n^2 + \frac{(K_2 b)^2}{2}\sin^2\phi_S^r\right)\frac{H_n^{(2)}(K_2 b\sin\phi_S^r)}{H_n^{(2)}(K_2 a\sin\phi_S^r)}$$

$$M[3,7] = M[3,8] = 0$$

$$M[4,1] = \mu_1\left((k_1 b\sin\phi_L^s)\frac{H_n^{(1)\prime}(k_1 b\sin\phi_L^s)}{H_n^{(1)}(k_1 b\sin\phi_L^s)} - n^2 + \frac{(K_1 b)^2}{2} - (k_1 b)^2\cos^2\phi_L^s\right)$$

$$M[4,2] = in\mu_1\left(1 - (K_1 b\sin\phi_S^s)\frac{H_n^{(1)\prime}(K_1 b\sin\phi_S^s)}{H_n^{(1)}(K_1 b\sin\phi_S^s)}\right)$$

$$M[4,3] = -\mu_2\left((k_2 b\sin\phi_L^r)\frac{H_n^{(1)\prime}(k_2 b\sin\phi_L^r)}{H_n^{(1)}(k_2 b\sin\phi_L^r)} - n^2 + \frac{(K_2 b)^2}{2} - (k_2 b)^2\cos^2\phi_L^r\right)\frac{H_n^{(1)}(k_2 b\sin\phi_L^r)}{H_n^{(1)}(k_2 a\sin\phi_L^r)}$$

$$M[4,4] = -\mu_2\left((k_2 b\sin\phi_L^r)\frac{H_n^{(2)\prime}(k_2 b\sin\phi_L^r)}{H_n^{(2)}(k_2 b\sin\phi_L^r)} - n^2 + \frac{(K_2 b)^2}{2} - (k_2 b)^2\cos^2\phi_L^r\right)\frac{H_n^{(2)}(k_2 b\sin\phi_L^r)}{H_n^{(2)}(k_2 a\sin\phi_L^r)}$$

$$M[4,5] = -in\mu_2\left(1 - (K_2 b\sin\phi_S^r)\frac{H_n^{(1)\prime}(K_2 b\sin\phi_S^r)}{H_n^{(1)}(K_2 b\sin\phi_S^r)}\right)\frac{H_n^{(1)}(K_2 b\sin\phi_S^r)}{H_n^{(1)}(K_2 a\sin\phi_S^r)}$$

$$M[4,6] = -in\mu_2\left(1 - (K_2 b\sin\phi_S^r)\frac{H_n^{(2)\prime}(K_2 b\sin\phi_S^r)}{H_n^{(2)}(K_2 b\sin\phi_S^r)}\right)\frac{H_n^{(2)}(K_2 b\sin\phi_S^r)}{H_n^{(2)}(K_2 a\sin\phi_S^r)}$$

$$M[4,7] = M[4,8] = M[5,1] = M[5,2] = 0$$

$$M[5,3] = -\mu_2\left((k_2 a\sin\phi_L^r)\frac{H_n^{(1)\prime}(k_2 a\sin\phi_L^r)}{H_n^{(1)}(k_2 a\sin\phi_L^r)} - n^2 + \frac{(K_2 a)^2}{2} - (k_2 a)^2\cos^2\phi_L^r\right)$$

$$M[5,4] = -\mu_2 \left((k_2 a \sin \phi_L^r) \frac{H_n^{(2)\prime}(k_2 a \sin \phi_L^r)}{H_n^{(2)}(k_2 a \sin \phi_L^r)} - n^2 + \frac{(K_2 a)^2}{2} \right.$$

$$\left. -(k_2 a)^2 \cos^2 \phi_L^r \right)$$

$$M[5,5] = -in\mu_2 \left(1 - (K_2 a \sin \phi_S^r) \frac{H_n^{(1)\prime}(K_2 a \sin \phi_S^r)}{H_n^{(1)}(K_2 a \sin \phi_S^r)} \right)$$

$$M[5,6] = -in\mu_2 \left(1 - (K_2 a \sin \phi_S^r) \frac{H_n^{(2)\prime}(K_2 a \sin \phi_S^r)}{H_n^{(2)}(K_2 a \sin \phi_S^r)} \right)$$

$$M[5,7] = in\mu_3 \left(1 - (K_3 a \sin \phi_S^t) \frac{J_n'(K_3 a \sin \phi_S^t)}{J_n(K_3 a \sin \phi_S^t)} \right)$$

$$M[5,8] = \mu_3 \left((k_3 a \sin \phi_L^t) \frac{J_n'(k_3 a \sin \phi_L^t)}{J_n(k_3 a \sin \phi_L^t)} - n^2 + \frac{(K_3 a)^2}{2} - (k_3 a)^2 \cos^2 \phi_L^t \right)$$

$$M[6,1] = M[6,2] = 0$$

$$M[6,3] = -in\mu_2 \left(1 - (k_2 a \sin \phi_L^r) \frac{H_n^{(1)\prime}(k_2 a \sin \phi_L^r)}{H_n^{(1)}(k_2 a \sin \phi_L^r)} \right)$$

$$M[6,4] = -in\mu_2 \left(1 - (k_2 a \sin \phi_L^r) \frac{H_n^{(2)\prime}(k_2 a \sin \phi_L^r)}{H_n^{(2)}(k_2 a \sin \phi_L^r)} \right)$$

$$M[6,5] = \mu_2 \left((K_2 a \sin \phi_S^r) \frac{H_n^{(1)\prime}(K_2 a \sin \phi_S^r)}{H_n^{(1)}(K_2 a \sin \phi_S^r)} - n^2 + \frac{(K_2 a)^2}{2} \sin^2 \phi_S^r \right)$$

$$M[6,6] = \mu_2 \left((K_2 a \sin \phi_S^r) \frac{H_n^{(2)\prime}(K_2 a \sin \phi_S^r)}{H_n^{(2)}(K_2 a \sin \phi_S^r)} - n^2 + \frac{(K_2 a)^2}{2} \sin^2 \phi_S^r \right)$$

$$M[6,7] = -\mu_3 \left((K_3 a \sin \phi_S^t) \frac{J_n'(K_3 a \sin \phi_S^t)}{J_n(K_3 a \sin \phi_S^t)} - n^2 + \frac{(K_3 a)^2}{2} \sin^2 \phi_S^t \right)$$

$$M[6,8] = in\mu_1 \left(1 - (k_3 a \sin \phi_L^t) \frac{J_n'(k_3 a \sin \phi_L^t)}{J_n(k_3 a \sin \phi_L^t)} \right)$$

$$M[7,1] = M[7,2] = 0$$

$$
\begin{aligned}
M[7,3] &= (k_2 a \sin\phi_L^r)\frac{H_n^{(1)\prime}(k_2 a \sin\phi_L^r)}{H_n^{(1)}(k_2 a \sin\phi_L^r)} \\
M[7,4] &= (k_2 a \sin\phi_L^r)\frac{H_n^{(2)\prime}(k_2 a \sin\phi_L^r)}{H_n^{(2)}(k_2 a \sin\phi_L^r)} \\
M[7,5] &= M[7,6] = in \\
M[7,7] &= -in \\
M[7,8] &= -(k_3 a \sin\phi_L^t)\frac{J_n'(k_3 a \sin\phi_L^t)}{J_n(k_3 a \sin\phi_L^t)} \\
M[8,1] &= M[8,2] = 0 \\
M[8,3] &= M[8,4] = in \\
M[8,5] &= -(K_2 a \sin\phi_S^r)\frac{H_n^{(1)\prime}(K_2 a \sin\phi_S^r)}{H_n^{(1)}(K_2 a \sin\phi_S^r)} \\
M[8,6] &= -(K_2 a \sin\phi_S^r)\frac{H_n^{(2)\prime}(K_2 a \sin\phi_S^r)}{H_n^{(2)}(K_2 a \sin\phi_S^r)} \\
M[8,7] &= (K_3 a \sin\phi_S^t)\frac{J_n'(K_3 a \sin\phi_S^t)}{J_n(K_3 a \sin\phi_S^t)} \\
M[8,8] &= -in \qquad (4.6)
\end{aligned}
$$

We are now left with expressing the b column vector, which represents the displacements and normal surface tractions due to the incident displacement field on the outer radius. The b column vector has the following components after factoring out the radial function common to all terms:

$$
\begin{aligned}
b[1] &= in \\
b[2] &= (k_1 b \sin \phi^i) \frac{H_n^{(1)\prime}(k_1 b \sin \phi^i)}{H_n^{(1)}(k_1 b \sin \phi^i)} \\
b[3] &= -in\mu_1 \left(1 - (k_1 b \sin \phi^i) \frac{H_n^{(1)\prime}(k_1 b \sin \phi^i)}{H_n^{(1)}(k_1 b \sin \phi^i)} \right) \\
b[4] &= -\mu_1 \left((k_1 b \sin \phi^i) \frac{H_n^{(1)\prime}(k_1 b \sin \phi^i)}{H_n^{(1)}(k_1 b \sin \phi^i)} - n^2 + \frac{(K_1 b)^2}{2} - (k_1 b)^2 \cos^2 \phi^i \right) \\
b[5] &= b[6] = b[7] = b[8] = 0 \qquad (4.7)
\end{aligned}
$$

The solution to the elastic scattering of incident displacement waves will be to write the modal coefficients of the displacement potentials in terms of the boundary conditions. The scattered displacement and stress fields can then be readily computed. The algebraic solution is given by the algebraic ratio of the determinants of the boundary value matrix and the substitution matrices via Cramer's rule. We define the determinants as $\Delta_i^{L,S,V}$, where $i = (0, 1, 2, .)$. The numbers in the expression for the determinant indicate the column of the matrix that b has been substituted into, and the superscript indicates the type of incident displacement field. For example, $x_1 = \frac{\Delta_1^L}{\Delta_0}$ is the coefficient of the displacement potential in the first column of the matrix. In this case, it is the modal coefficient of the scattered longitudinal displacement field due to an incident longitudinal displacement field. The displacement potentials of interest in the scattering problem are the scattered ones and can be written in terms of the determinants as:

$$
\begin{aligned}
a_{1n} &= \frac{J_n(k_1 b \sin\phi^i)}{H_n^{(1)}(k_1 b \sin\phi_L^s)} \left(\frac{\Delta_1^L}{\Delta_0}\right) \\
b_{1n} &= \left(\frac{K_1}{k_1}\right) \frac{J_n(k_1 b \sin\phi^i)}{H_n^{(1)}(K_1 b \sin\phi_S^s)} \left(\frac{\Delta_2^L}{\Delta_0}\right)
\end{aligned}
\tag{4.8}
$$

where the factors in front of the potentials are the ratio of terms factored out of the boundary condition matrix prior to taking the determinants.

4.2.1 Scattered fields

The scattered fields created from an incident longitudinal displacement field will be of both the longitudinal and transverse types. While longitudinal and transverse displacement fields do not interact with each other in the external fields, they are coupled together through interfaces. The displacement potentials of the scattered fields are given, upon substitution of the expressions for the coefficients, as:

$$
\pi_L^s = \sum_{n=0}^{\infty} \frac{J_n(k_1 b \sin\phi^i)}{H_n^{(1)}(k_1 b \sin\phi_L^s)} \left(\frac{\Delta_1^L}{\Delta_0}\right) i^n H_n^{(1)}(k_1 r \sin\phi_L^s) e^{in\theta} e^{ik_1 z \cos\phi_L^s} \tag{4.9}
$$

$$
\pi_S^s = \sum_{n=0}^{\infty} \left(\frac{K_1}{k_1}\right) \frac{J_n(k_1 b \sin\phi^i)}{H_n^{(1)}(K_1 b \sin\phi_S^s)} \left(\frac{\Delta_2^L}{\Delta_0}\right) i^n H_n^{(1)}(K_1 r \sin\phi_S^s) e^{in\theta} e^{iK_1 z \cos\phi_S^s}
$$

We can see from the displacement potentials that the coupling of these scattered displacement fields to the displacement fields inside of the interface layer and in the interior are completely contained within the determinant ratios. All other factors in the displacement potentials are dependent only on the external field variables, including the angle of propagation. If we take a look at the expression for Δ_0 (two pages forward), as complicated as it appears, there is a clear structure to the algebraic solution that can be seen. First we should note that the subscripts on the components of the factors in the determinant actually represent the matrix element position after extensive simplification of the boundary condition matrix. Secondly, we note that through careful manipulation, we have not mixed the expressions for the interior field with those of the exterior field in the matrix manipulations. The first four terms in Δ_0 are coupling terms between the exterior field and interface layer, and the remaining four terms are coupling between the interface layer and the interior. This coupling through the interface makes logical sense, since the exterior and interior fields are separated by the interface layer, the dependence of interior and exterior fields must be through the properties of the interface layer. The expressions for $\Delta_0, \Delta_1^L, \Delta_2^L$ are given as follows, with the last four elements of the determinant identical for all three determinants, as expected, since the relations between the interface layer and interior region must not be affected by substitution of the column vector b into the columns corresponding to components from the scattered field. The expressions for the determinants are:

$$\Delta_0 = D_0\left[1 - C_{11}C_{88} - C_{12}C_{78} - C_{21}C_{87} - C_{22}C_{77} + (C_{11}C_{22} - C_{21}C_{21})(C_{77}C_{88} - C_{87}C_{78})\right] \tag{4.10}$$

$$C_{11} = \frac{1}{\left(F_{b2}^T - \tilde{H}_{b2}^T\right)\frac{H_n^{(2)}(\beta_2 b)}{H_n^{(2)}(\beta_2 a)}}\left[F_{b2}^T + \frac{\frac{(\beta_2 b)^2\mu_2}{2(\mu_2-\mu_1)}}{(1 - H_{b1}^L)} + \frac{\left(\frac{(\beta_2 b)^2\mu_2}{2(\mu_2-\mu_1)}\frac{(\beta_2^2-\sigma^2)b^2\mu_2}{2(\mu_2-\mu_1)}\right)\left(n^2 - H_{b1}^L H_{b1}^T + \frac{(\beta_1 b)^2\mu_1}{2(\mu_2-\mu_1)}\right)\left(n^2 - H_{b1}^L + \frac{(\beta_1^2-\sigma^2)b^2\mu_1}{2(\mu_2-\mu_1)}\right)}{(1 - H_{b1}^L)\,D_0}\right]$$

$$C_{12} = \frac{1}{\left(F_{b2}^T - \tilde{H}_{b2}^T\right)\frac{H_n^{(2)}(\beta_2 b)}{H_n^{(2)}(\beta_2 a)}}\left[1 + \frac{\left(\frac{(\beta_2^2-\sigma^2)b^2\mu_2}{2(\mu_2-\mu_1)}\right)\left(n^2 - H_{b1}^L H_{b1}^T + \frac{(\beta_1 b)^2\mu_1}{2(\mu_2-\mu_1)}\right)}{D_0}\right]$$

$$C_{21} = \frac{1}{\left(F_{b2}^L - \tilde{H}_{b2}^L\right)\frac{H_n^{(2)}(\alpha_2 b)}{H_n^{(2)}(\alpha_2 a)}}\left[n^2 + \frac{H_{b1}^L\frac{(\beta_2 b)^2\mu_2}{2(\mu_2-\mu_1)}}{(1 - H_{b1}^L)} + \frac{\left(\frac{(\beta_2 b)^2\mu_2}{2(\mu_2-\mu_1)}\frac{(\beta_2^2-\sigma^2)b^2\mu_2}{2(\mu_2-\mu_1)}\right)\left(n^2 - H_{b1}^L H_{b1}^T + H_{b1}^L\frac{(\beta_1 b)^2\mu_1}{2(\mu_2-\mu_1)}\right)\left(n^2 - H_{b1}^L + \frac{(\beta_1^2-\sigma^2)b^2\mu_1}{2(\mu_2-\mu_1)}\right)}{(1 - H_{b1}^L)\,D_0}\right]$$

$$C_{22} = \frac{1}{\left(F_{b2}^L - \tilde{H}_{b2}^L\right)\frac{H_n^{(2)}(\alpha_2 b)}{H_n^{(2)}(\alpha_2 a)}}\left[F_{b2}^L + \frac{\left(\frac{(\beta_2^2-\sigma^2)b^2\mu_2}{2(\mu_2-\mu_1)}\right)\left(n^2 - H_{b1}^L H_{b1}^T + H_{b1}^L\frac{(\beta_1 b)^2\mu_1}{2(\mu_2-\mu_1)}\right)}{D_0}\right]$$

$$C_{77} = \frac{-1}{F_{a2}^L}\left[\tilde{H}_{a2}^L - F_{a2}^L\frac{H_n^{(2)}(\alpha_2 b)}{H_n^{(2)}(\alpha_2 a)} + \frac{\left(\frac{(\beta_2^2-\sigma^2)a^2\mu_2}{2(\mu_2-\mu_3)}\right)\left(n^2 - J_{a3}^L J_{a3}^T + J_{a3}^L\frac{(\beta_3 a)^2\mu_3}{2(\mu_2-\mu_3)}\right)}{M_0}\right]$$

$$C_{78} = \frac{-1}{F_{a2}^L}\left[n^2 + \frac{J_{a3}^L\frac{(\beta_2 a)^2\mu_2}{2(\mu_2-\mu_3)}}{(1 - J_{a3}^L)} + \frac{\left(\frac{(\beta_2 a)^2\mu_2}{2(\mu_2-\mu_3)}\frac{(\beta_2^2-\sigma^2)a^2\mu_2}{2(\mu_2-\mu_3)}\right)\left(n^2 - J_{a3}^L J_{a3}^T + J_{a3}^L\frac{(\beta_3 a)^2\mu_3}{2(\mu_2-\mu_3)}\right)\left(n^2 - J_{a3}^L + \frac{(\beta_3^2-\sigma^2)a^2\mu_3}{2(\mu_2-\mu_3)}\right)}{(1 - J_{a3}^L)\,M_0}\right]$$

$$C_{87} = \frac{-1}{F_{a2}^T}\left[1 + \frac{\left(\frac{(\beta_2^2-\sigma^2)a^2\mu_2}{2(\mu_2-\mu_3)}\right)\left(n^2 - J_{a3}^L J_{a3}^T + \frac{(\beta_3 a)^2\mu_3}{2(\mu_2-\mu_3)}\right)}{M_0}\right]$$

$$C_{88} = \frac{-1}{F_{a2}^T}\left[\left(\tilde{H}_{a2}^T - F_{a2}^T\frac{H_n^{(2)}(\beta_2 b)}{H_n^{(2)}(\beta_2 a)}\right) + \frac{\frac{(\beta_2 a)^2\mu_2}{2(\mu_2-\mu_3)}}{(1 - J_{a3}^L)} + \frac{\left(\frac{(\beta_2 a)^2\mu_2}{2(\mu_2-\mu_3)}\frac{(\beta_2^2-\sigma^2)a^2\mu_2}{2(\mu_2-\mu_3)}\right)\left(n^2 - J_{a3}^T J_{a3}^L + \frac{(\beta_3^2-\sigma^2)a^2\mu_3}{2(\mu_2-\mu_3)}\right)\left(n^2 - J_{a3}^L + \frac{(\beta_3^2-\sigma^2)a^2\mu_3}{2(\mu_2-\mu_3)}\right)}{(1 - J_{a3}^L)\,M_0}\right]$$

The following define the expressions in $C_{11,12,..}$.

$$D_0 = n^2\left(1-H^L_{b1}\right)\left(1-H^T_{b1}\right) - \left(n^2 - H^T_{b1} + \frac{(\beta_1 b)^2\mu_1}{2(\mu_2-\mu_1)}\right)\left(n^2 - H^L_{b1} + \frac{(\beta_1^2-\sigma^2)\,b^2\mu_1}{2(\mu_2-\mu_1)}\right)$$

$$M_0 = n^2\left(1-J^L_{a3}\right)\left(1-J^T_{a3}\right) - \left(n^2 - J^T_{a3} + \frac{(\beta_3 a)^2\mu_3}{2(\mu_2-\mu_3)}\right)\left(n^2 - J^L_{a3} + \frac{(\beta_3^2-\sigma^2)\,a^2\mu_3}{2(\mu_2-\mu_3)}\right)$$

$$H^L_{b1} = \frac{(\alpha_1 b)H_n^{(1)\prime}(\alpha_1 b)}{H_n^{(1)}(\alpha_1 b)} \qquad H^T_{b1} = \frac{(\beta_1 b)H_n^{(1)\prime}(\beta_1 b)}{H_n^{(1)}(\beta_1 b)} \qquad H^L_{b2} = \frac{(\alpha_2 b)H_n^{(1)\prime}(\alpha_2 b)}{H_n^{(1)}(\alpha_2 b)} \qquad H^T_{b2} = \frac{(\beta_2 b)H_n^{(1)\prime}(\beta_2 b)}{H_n^{(1)}(\beta_2 b)}$$

$$\tilde{H}^L_{b2} = \frac{(\alpha_2 b)H_n^{(2)\prime}(\alpha_2 b)}{H_n^{(2)}(\alpha_2 b)} \qquad \tilde{H}^T_{b2} = \frac{(\beta_2 b)H_n^{(2)\prime}(\beta_2 b)}{H_n^{(2)}(\beta_2 b)} \qquad \tilde{H}^L_{a2} = \frac{(\alpha_2 a)H_n^{(2)\prime}(\alpha_2 a)}{H_n^{(2)}(\alpha_2 a)} \qquad \tilde{H}^T_{a2} = \frac{(\beta_2 a)H_n^{(2)\prime}(\beta_2 a)}{H_n^{(2)}(\beta_2 a)}$$

$$J^L_{b1} = \frac{(\alpha_1 b)J_n'(\alpha_1 b)}{J_n(\alpha_1 b)} \qquad J^T_{b1} = \frac{(\beta_1 b)J_n'(\beta_1 b)}{J_n(\beta_1 b)} \qquad J^L_{a3} = \frac{(\alpha_3 a)J_n'(\alpha_3 a)}{J_n(\alpha_3 a)} \qquad J^T_{a3} = \frac{(\beta_3 a)J_n'(\beta_3 a)}{J_n(\beta_3 a)}$$

$$F^L_{a2} = (\alpha_2 a)\left[\frac{H_n^{(1)\prime}(\alpha_2 a)H_n^{(2)}(\alpha_2 a) - H_n^{(2)\prime}(\alpha_2 a)H_n^{(1)}(\alpha_2 a)}{H_n^{(1)}(\alpha_2 b)H_n^{(2)}(\alpha_2 a) - H_n^{(2)}(\alpha_2 b)H_n^{(1)}(\alpha_2 a)}\right] \qquad F^L_{b2} = (\alpha_2 b)\left[\frac{H_n^{(1)\prime}(\alpha_2 b)H_n^{(2)}(\alpha_2 a) - H_n^{(2)\prime}(\alpha_2 b)H_n^{(1)}(\alpha_2 a)}{H_n^{(1)}(\alpha_2 b)H_n^{(2)}(\alpha_2 a) - H_n^{(2)}(\alpha_2 b)H_n^{(1)}(\alpha_2 a)}\right]$$

$$F^T_{a2} = (\beta_2 a)\left[\frac{H_n^{(1)\prime}(\beta_2 a)H_n^{(2)}(\beta_2 a) - H_n^{(2)\prime}(\beta_2 a)H_n^{(1)}(\beta_2 a)}{H_n^{(1)}(\beta_2 b)H_n^{(2)}(\beta_2 a) - H_n^{(2)}(\beta_2 b)H_n^{(1)}(\beta_2 a)}\right] \qquad F^T_{b2} = (\beta_2 b)\left[\frac{H_n^{(1)\prime}(\beta_2 b)H_n^{(2)}(\beta_2 a) - H_n^{(2)\prime}(\beta_2 b)H_n^{(1)}(\beta_2 a)}{H_n^{(1)}(\beta_2 b)H_n^{(2)}(\beta_2 a) - H_n^{(2)}(\beta_2 b)H_n^{(1)}(\beta_2 a)}\right]$$

$$\alpha_{1,2,3} = k_{1,2,3}\sin\phi_{1,3,5} \qquad \beta_{1,2,3} = K_{1,2,3}\sin\phi_{2,4,6}$$

$$\sigma = k_{1,2,3}\cos\phi_{1,3,5} = K_{1,2,3}\cos\phi_{2,4,6}$$

$$\Delta_1^L = D_1^L \left[1 - C_{11}C_{88} - C_{12}C_{78} - C_{21}C_{87} - C_{22}C_{77} + (C_{11}C_{22} - C_{21}C_{21})(C_{77}C_{88} - C_{87}C_{78})\right] \tag{4.11}$$

$$C_{11} = \frac{1}{\left(F_{b2}^T - \tilde{H}_{b2}^T\right)\frac{H_n^{(2)}(\beta_2 b)}{H_n^{(2)}(\beta_2 a)}}\left[F_{b2}^T + \frac{\frac{(\beta_2 b)^2\mu_2}{2(\mu_2-\mu_1)}}{\left(1 - J_{b1}^L\right)} + \frac{\left(\frac{(\beta_2 b)^2\mu_2}{2(\mu_2-\mu_1)}\frac{\left(\beta_2^2-\sigma^2\right)b^2\mu_2}{2(\mu_2-\mu_1)}\right)\left(n^2 - J_{b1}^L H_{b1}^T + \frac{(\beta_1 b)^2\mu_1}{2(\mu_2-\mu_1)}\right)\left(n^2 - J_{b1}^L + \frac{\left(\beta_1^2-\sigma^2\right)b^2\mu_1}{2(\mu_2-\mu_1)}\right)}{\left(1 - J_{b1}^L\right)D_1^L}\right]$$

$$C_{12} = \frac{1}{\left(F_{b2}^T - \tilde{H}_{b2}^T\right)\frac{H_n^{(2)}(\beta_2 b)}{H_n^{(2)}(\beta_2 a)}}\left[1 + \frac{\left(\frac{\left(\beta_2^2-\sigma^2\right)b^2\mu_2}{2(\mu_2-\mu_1)}\right)\left(n^2 - J_{b1}^L H_{b1}^T + \frac{(\beta_1 b)^2\mu_1}{2(\mu_2-\mu_1)}\right)}{D_1^L}\right]$$

$$C_{21} = \frac{1}{\left(F_{b2}^L - \tilde{H}_{b2}^L\right)\frac{H_n^{(2)}(\alpha_2 b)}{H_n^{(2)}(\alpha_2 a)}}\left[n^2 + \frac{J_{b1}^L\frac{(\beta_2 b)^2\mu_2}{2(\mu_2-\mu_1)}}{\left(1 - J_{b1}^L\right)} + \frac{\left(\frac{(\beta_2 b)^2\mu_2}{2(\mu_2-\mu_1)}\frac{\left(\beta_2^2-\sigma^2\right)b^2\mu_2}{2(\mu_2-\mu_1)}\right)\left(n^2 - J_{b1}^L H_{b1}^T + J_{b1}^L\frac{(\beta_1 b)^2\mu_1}{2(\mu_2-\mu_1)}\right)\left(n^2 - J_{b1}^L + \frac{\left(\beta_1^2-\sigma^2\right)b^2\mu_1}{2(\mu_2-\mu_1)}\right)}{\left(1 - J_{b1}^L\right)D_1^L}\right]$$

$$C_{22} = \frac{1}{\left(F_{b2}^L - \tilde{H}_{b2}^L\right)\frac{H_n^{(2)}(\alpha_2 b)}{H_n^{(2)}(\alpha_2 a)}}\left[F_{b2}^L + \frac{\left(\frac{\left(\beta_2^2-\sigma^2\right)b^2\mu_2}{2(\mu_2-\mu_1)}\right)\left(n^2 - J_{b1}^L H_{b1}^T + J_{b1}^L\frac{(\beta_1 b)^2\mu_1}{2(\mu_2-\mu_1)}\right)}{D_1^L}\right]$$

$$D_1^L = n^2\left(1 - J_{b1}^L\right)\left(1 - H_{b1}^T\right) - \left(n^2 - H_{b1}^T + \frac{(\beta_1 b)^2\mu_1}{2(\mu_2 - \mu_1)}\right)\left(n^2 - J_{b1}^L + \frac{\left(\beta_1^2 - \sigma^2\right)b^2\mu_1}{2(\mu_2 - \mu_1)}\right)$$

$$\Delta_2^L = D_2^L\left[1 - C_{11}C_{88} - C_{12}C_{78} - C_{21}C_{87} - C_{22}C_{77} + (C_{11}C_{22} - C_{21}C_{21})(C_{77}C_{88} - C_{87}C_{78})\right] \tag{4.12}$$

$$C_{11} = \frac{1}{\left(F_{b2}^T - \tilde{H}_{b2}^T\right)\frac{H_n^{(2)}(\beta_2 b)}{H_n^{(2)}(\beta_2 a)}}\left[F_{b2}^T + \frac{\frac{(\beta_2 b)^2\mu_2}{2(\mu_2-\mu_1)}}{(1-H_{b1}^L)} - \frac{\left(\frac{(\beta_2 b)^2\mu_2}{2(\mu_2-\mu_1)}\frac{(\beta_2^2-\sigma^2)b^2\mu_2}{2(\mu_2-\mu_1)}\right)\left(n^2 - H_{b1}^L + \frac{(\beta_1^2-\sigma^2)b^2\mu_1}{2(\mu_2-\mu_1)}\right)}{(1-H_{b1}^L)\left(n^2 - 1 - \frac{(\beta_1^2-\sigma^2)b^2\mu_1}{2(\mu_2-\mu_1)}\right)}\right]$$

$$C_{12} = \frac{1}{\left(F_{b2}^T - \tilde{H}_{b2}^T\right)\frac{H_n^{(2)}(\beta_2 b)}{H_n^{(2)}(\beta_2 a)}}\left[1 + \frac{\left(\frac{(\beta_2^2-\sigma^2)b^2\mu_2}{2(\mu_2-\mu_1)}\right)}{\left(n^2 - 1 + \frac{(\beta_2^2-\sigma^2)b^2\mu_2}{2(\mu_2-\mu_1)}\right)}\right]$$

$$C_{21} = \frac{1}{\left(F_{b2}^L - \tilde{H}_{b2}^L\right)\frac{H_n^{(2)}(\alpha_2 b)}{H_n^{(2)}(\alpha_2 a)}}\left[n^2 + \frac{H_{b1}^L\frac{(\beta_2 b)^2\mu_2}{2(\mu_2-\mu_1)}}{(1-H_{b1}^L)} - \frac{\left(\frac{(\beta_2 b)^2\mu_2}{2(\mu_2-\mu_1)}\frac{(\beta_2^2-\sigma^2)b^2\mu_2}{2(\mu_2-\mu_1)}\right)\left(n^2 - H_{b1}^L + \frac{(\beta_1^2-\sigma^2)b^2\mu_1}{2(\mu_2-\mu_1)}\right)}{\left(n^2 - 1 + \frac{(\beta_1^2-\sigma^2)b^2\mu_1}{2(\mu_2-\mu_1)}\right)}\right]$$

$$C_{22} = \frac{1}{\left(F_{b2}^L - \tilde{H}_{b2}^L\right)\frac{H_n^{(2)}(\alpha_2 b)}{H_n^{(2)}(\alpha_2 a)}}\left[F_{b2}^L + \frac{\left(\frac{(\beta_2^2-\sigma^2)b^2\mu_2}{2(\mu_2-\mu_1)}\right)}{\left(n^2 - 1 + \frac{(\beta_1^2-\sigma^2)b^2\mu_1}{2(\mu_2-\mu_1)}\right)}\right]$$

$$D_2^L = n\left(H_{b1}^L - J_{b1}^L\right)\left(n^2 - 1 + \frac{(\beta_1^2-\sigma^2)\,b^2\mu_1}{2(\mu_2-\mu_1)}\right)$$

4.2.2 Dynamic stress concentrations

The expressions for the scattered displacement potentials allow us to explicitly write the expressions for the dynamic stress concentrations. These can then be evaluated for varying elastic properties and geometry to determine the effect of an interface layer on the dynamic stresses. We normalize the dynamic stress concentrations by calculating the incident stress field and computing σ^*, the dynamic stress concentration [78]. There will be three terms contributing to σ^* and it is preferable to review the individual contributions in order to interpret the physics of the problem. Since the incident longitudinal displacement field will only produce shear stresses when the angle of incidence is nonzero, we will normalize the scattered shear stresses by the incident longitudinal stress. The expressions for stress concentration are then given as follows:

$$\sigma^*_{rr} = \frac{\left|\sigma^L_{rr} + \sigma^S_{rr}\right|}{\left|\sigma^i_{xx}\right|}$$

$$\sigma^L_{rr} = \sum_{n=0}^{\infty} \frac{2\mu_1 i}{k_1 r^2} \left((k_1 r \sin\phi^s_L) \frac{H^{(1)\prime}_n(k_1 r \sin\phi^s_L)}{H^{(1)}_n(k_1 r \sin\phi^s_L)} - n^2 + \frac{(K_1 r)^2}{2} - (k_1 r \cos\phi^s_L)^2 \right)$$

$$\frac{J_n(k_1 b \sin\phi^i)}{H^{(1)}_n(k_1 b \sin\phi^s_L)} \left(\frac{\Delta^L_1}{\Delta_0}\right) i^n H^{(1)}_n(k_1 r \sin\phi^s_L) e^{in\theta} e^{ik_1 z \cos\phi^s_L}$$

$$\sigma^S_{rr} = \sum_{n=0}^{\infty} -n \frac{2\mu_1}{K_1 r^2} \left(1 - (K_1 r \sin\phi^s_S) \frac{H^{(1)\prime}_n(K_1 r \sin\phi^s_S)}{H^{(1)}_n(K_1 r \sin\phi^s_S)} \right)$$

$$\left(\frac{K_1}{k_1}\right) \frac{J_n(k_1 b \sin\phi^i)}{H^{(1)}_n(K_1 b \sin\phi^s_S)} \left(\frac{\Delta^L_2}{\Delta_0}\right) i^n H^{(1)}_n(K_1 r \sin\phi^s_S) e^{in\theta} e^{iK_1 z \cos\phi^s_S} \qquad (4.13)$$

and the shearing stress concentrations are:

$$\sigma^*_{r\theta} = \frac{\left|\sigma^L_{r\theta} + \sigma^S_{r\theta}\right|}{\left|\sigma^i_{xx}\right|}$$

$$\sigma^L_{r\theta} = \sum_{n=0}^{\infty} -n\frac{2\mu_1}{k_1 r^2}\left(1-(k_1 r\sin\phi^s_L)\frac{H^{(1)\prime}_n(k_1 r\sin\phi^s_L)}{H^{(1)}_n(k_1 r\sin\phi^s_L)}\right)$$

$$\frac{J_n(k_1 b\sin\phi^i)}{H^{(1)}_n(k_1 b\sin\phi^s_L)}\left(\frac{\Delta^L_1}{\Delta_0}\right) i^n H^{(1)}_n(k_1 r\sin\phi^s_L)e^{in\theta}e^{ik_1 z\cos\phi^s_L}$$

$$\sigma^S_{r\theta} = \sum_{n=0}^{\infty} -\frac{2\mu_1 i}{K_1 r^2}\left((K_1 r\sin\phi^s_S)\frac{H^{(1)\prime}_n(K_1 r\sin\phi^s_S)}{H^{(1)}_n(K_1 r\sin\phi^s_S)} - n^2 + \frac{(K_1 r)^2}{2}\cos^2\phi^s_S\right)$$

$$\left(\frac{K_1}{k_1}\right)\frac{J_n(k_1 b\sin\phi^i)}{H^{(1)}_n(K_1 b\sin\phi^s_S)}\left(\frac{\Delta^L_2}{\Delta_0}\right) i^n H^{(1)}_n(K_1 r\sin\phi^s_S)e^{in\theta}e^{iK_1 z\cos\phi^s_S} \tag{4.14}$$

and finally the hoop stress concentrations can be written as:

$$\sigma^*_{\theta\theta} = \frac{\left|\sigma^L_{\theta\theta}+\sigma^S_{\theta\theta}\right|}{\left|\sigma^i_{xx}\right|}$$

$$\sigma^L_{\theta\theta} = \sum_{n=0}^{\infty} \frac{2\mu_1}{k_1 r^2}\left(\frac{(K_1 r)^2}{2} - (k_1 r)^2 + n^2 - (k_1 r\sin\phi^s_L)\frac{H^{(1)\prime}_n(k_1 r\sin\phi^s_L)}{H^{(1)}_n(k_1 r\sin\phi^s_L)}\right)$$

$$\frac{J_n(k_1 b\sin\phi^i)}{H^{(1)}_n(k_1 b\sin\phi^s_L)}\left(\frac{\Delta^L_1}{\Delta_0}\right) i^n H^{(1)}_n(k_1 r\sin\phi^s_L)e^{in\theta}e^{ik_1 z\cos\phi^s_L}$$

$$\sigma^S_{\theta\theta} = \sum_{n=0}^{\infty} in\frac{2\mu_1}{K_1 r^2}\left(1-(K_1 r\sin\phi^s_S)\frac{H^{(1)\prime}_n(K_1 r\sin\phi^s_S)}{H^{(1)}_n(K_1 r\sin\phi^s_S)}\right)$$

$$\left(\frac{K_1}{k_1}\right)\frac{J_n(k_1 b\sin\phi^i)}{H^{(1)}_n(K_1 b\sin\phi^s_S)}\left(\frac{\Delta^L_2}{\Delta_0}\right) i^n H^{(1)}_n(K_1 r\sin\phi^s_S)e^{in\theta}e^{iK_1 z\cos\phi^s_S} \tag{4.15}$$

4.3 Incident coupled transverse wave

In the case of an incident transverse displacement wave of the π_S type, we have for the incident displacement potential

$$\pi^i_S = \sum_{n=0}^{\infty} \epsilon_n i^n J_n(K_1 r\sin\phi^i)e^{in\theta}e^{iK_1 z\cos\phi^i} \tag{4.16}$$

The same boundary conditions are needed to solve for the scattered fields as in the incident longitudinal case. We will need to solve the 8×8 system to calculate the modal coefficients of the displacement potentials. The matrix M is unchanged, but the column vector b, which represents the incident field components of displacement and normal surface traction on the outer boundary, is modified. The b column vector for the incident transverse wave has the components:

$$\begin{aligned}
b[1] &= -(K_1 b \sin\phi^i)\frac{H_n^{(1)\prime}(K_1 b \sin\phi^i)}{H_n^{(1)}(K_1 b \sin\phi^i)} \\
b[2] &= in \\
b[3] &= \mu_1\left((K_1 b \sin\phi^i)\frac{H_n^{(1)\prime}(K_1 b \sin\phi^i)}{H_n^{(1)}(K_1 b \sin\phi^i)} - n^2 + \frac{(K_1 b)^2}{2}\sin^2\phi^i\right) \\
b[4] &= -in\mu_1\left(1 - (K_1 b \sin\phi^i)\frac{H_n^{(1)\prime}(K_1 b \sin\phi^i)}{H_n^{(1)}(K_1 b \sin\phi^i)}\right) \\
b[5] &= b[6] = b[7] = b[8] = 0 \qquad (4.17)
\end{aligned}$$

We have factored out the radial functions in the incident field. We now substitute the column vector b into the matrix M and recompute the determinants, solving for the unknown modal coefficients A_{1n} and B_{1n}. The expression for Δ_0 remains unchanged from the previously given expression and the coefficients of the scattered fields can be written as:

$$A_{1n} = \left(\frac{k_1}{K_1}\right)\frac{J_n(K_1 b \sin\phi^i)}{H_n^{(1)}(k_1 b \sin\phi_L^s)}\left(\frac{\Delta_1^S}{\Delta_0}\right)$$

$$B_{1n} = \frac{J_n(K_1 b \sin\phi^i)}{H_n^{(1)}(K_1 b \sin\phi_S^s)}\left(\frac{\Delta_2^S}{\Delta_0}\right) \tag{4.18}$$

where the factors in front of the potentials are the ratio of terms factored out of the boundary condition matrix prior to taking the determinants.

4.3.1 Scattered fields

The scattered fields due to the incident transverse displacement field will be both longitudinal and transverse. The displacement potentials are given, upon substitution of the expressions for the coefficients, as:

$$\pi_L^s = \sum_{n=0}^{\infty}\left(\frac{k_1}{K_1}\right)\frac{J_n(K_1 b \sin\phi^i)}{H_n^{(1)}(k_1 b \sin\phi_L^s)}\left(\frac{\Delta_1^S}{\Delta_0}\right) i^n H_n^{(1)}(k_1 r \sin\phi_L^s) e^{in\theta} e^{ik_1 z \cos\phi_L^s}$$

$$\pi_S^s = \sum_{n=0}^{\infty}\frac{J_n(K_1 b \sin\phi^i)}{H_n^{(1)}(K_1 b \sin\phi_S^s)}\left(\frac{\Delta_2^S}{\Delta_0}\right) i^n H_n^{(1)}(K_1 r \sin\phi_S^s) e^{in\theta} e^{iK_1 z \cos\phi_S^s} \tag{4.19}$$

with the expressions for Δ_1^S, Δ_2^S given by:

$$\Delta_1^S = D_1^S\left[1 - C_{11}C_{88} - C_{12}C_{78} - C_{21}C_{87} - C_{22}C_{77} + (C_{11}C_{22} - C_{21}C_{21})(C_{77}C_{88} - C_{87}C_{78})\right] \tag{4.19}$$

$$C_{11} = \frac{1}{\left(F_{b2}^T - \tilde{H}_{b2}^T\right)\frac{H_n^{(2)}(\beta_2 b)}{H_n^{(2)}(\beta_2 a)}}\left[F_{b2}^T - \frac{J_{b1}^T\frac{(\beta_2 b)^2\mu_2}{2(\mu_2-\mu_1)}}{\left(n^2 - J_{b1}^T + \frac{(\beta_1 b)^2\mu_1}{2(\mu_2-\mu_1)}\right)} + \frac{\left(\frac{(\beta_2 b)^2\mu_2}{2(\mu_2-\mu_1)}\frac{(\beta_2^2-\sigma^2)b^2\mu_2}{2(\mu_2-\mu_1)}\right)\left(1 - J_{b1}^T\right)\left(n^2 + \frac{(\beta_1 b)^2\mu_1}{2(\mu_2-\mu_1)}\right)}{\left(n^2 - 1 + \frac{(\beta_1 b)^2\mu_1}{2(\mu_2-\mu_1)}\right)}\right]$$

$$C_{12} = \frac{1}{\left(F_{b2}^T - \tilde{H}_{b2}^T\right)\frac{H_n^{(2)}(\beta_2 b)}{H_n^{(2)}(\beta_2 a)}}\left[n + \frac{\left(\frac{(\beta_2^2-\sigma^2)b^2\mu_2}{2(\mu_2-\mu_1)}\right)\left(n^2 + \frac{(\beta_1 b)^2\mu_1}{2(\mu_2-\mu_1)}\right)}{n\left(n^2 - 1 + \frac{(\beta_1 b)^2\mu_1}{2(\mu_2-\mu_1)}\right)}\right]$$

$$C_{21} = \frac{n}{\left(F_{b2}^L - \tilde{H}_{b2}^L\right)\frac{H_n^{(2)}(\alpha_2 b)}{H_n^{(2)}(\alpha_2 a)}}\left[1 + \frac{\frac{(\beta_2 b)^2\mu_2}{2(\mu_2-\mu_1)}}{\left(n^2 - J_{b1}^T + \frac{(\beta_1 b)^2\mu_1}{2(\mu_2-\mu_1)}\right)} - \frac{\left(\frac{(\beta_2 b)^2\mu_2}{2(\mu_2-\mu_1)}\frac{(\beta_2^2-\sigma^2)b^2\mu_2}{2(\mu_2-\mu_1)}\right)\left(1 - J_{b1}^T\right)}{\left(n^2 - J_{b1}^T + \frac{(\beta_1 b)^2\mu_1}{2(\mu_2-\mu_1)}\right)\left(n^2 - 1 + \frac{(\beta_1 b)^2\mu_1}{2(\mu_2-\mu_1)}\right)}\right]$$

$$C_{22} = \frac{1}{\left(F_{b2}^L - \tilde{H}_{b2}^L\right)\frac{H_n^{(2)}(\alpha_2 b)}{H_n^{(2)}(\alpha_2 a)}}\left[F_{b2}^L - \frac{\left(\frac{(\beta_2^2-\sigma^2)b^2\mu_2}{2(\mu_2-\mu_1)}\right)}{\left(n^2 - 1 + \frac{(\beta_1 b)^2\mu_1}{2(\mu_2-\mu_1)}\right)}\right]$$

$$D_1^S = n\left(H_{b1}^T - J_{b1}^T\right)\left(n^2 - 1 + \frac{(\beta_1 b)^2\mu_1}{2(\mu_2 - \mu_1)}\right)$$

$$\Delta_2^S = D_2^S\left[1 - C_{11}C_{88} - C_{12}C_{78} - C_{21}C_{87} - C_{22}C_{77} + (C_{11}C_{22} - C_{21}C_{21})(C_{77}C_{88} - C_{87}C_{78})\right] \tag{4.20}$$

$$C_{11} = \frac{1}{\left(F_{b2}^T - \tilde{H}_{b2}^T\right)\frac{H_n^{(2)}(\beta_2 b)}{H_n^{(2)}(\beta_2 a)}}\left[F_{b2}^T + \frac{\frac{(\beta_2 b)^2\mu_2}{2(\mu_2-\mu_1)}}{\left(1-H_{b1}^L\right)} + \frac{\left(\frac{(\beta_2 b)^2\mu_2}{2(\mu_2-\mu_1)}\frac{(\beta_2^2-\sigma^2)b^2\mu_2}{2(\mu_2-\mu_1)}\right)\left(n^2 - H_{b1}^L J_{b1}^T + \frac{(\beta_1 b)^2\mu_1}{2(\mu_2-\mu_1)}\right)\left(n^2 - H_{b1}^L + \frac{(\beta_1^2-\sigma^2)b^2\mu_1}{2(\mu_2-\mu_1)}\right)}{\left(1-H_{b1}^L\right)D_2^S}\right]$$

$$C_{12} = \frac{1}{\left(F_{b2}^T - \tilde{H}_{b2}^T\right)\frac{H_n^{(2)}(\beta_2 b)}{H_n^{(2)}(\beta_2 a)}}\left[1 + \frac{\left(\frac{(\beta_2^2-\sigma^2)b^2\mu_2}{2(\mu_2-\mu_1)}\right)\left(n^2 - H_{b1}^L J_{b1}^T + \frac{(\beta_1 b)^2\mu_1}{2(\mu_2-\mu_1)}\right)}{D_2^S}\right]$$

$$C_{21} = \frac{1}{\left(F_{b2}^L - \tilde{H}_{b2}^L\right)\frac{H_n^{(2)}(\alpha_2 b)}{H_n^{(2)}(\alpha_2 a)}}\left[n^2 + \frac{H_{b1}^L\frac{(\beta_2 b)^2\mu_2}{2(\mu_2-\mu_1)}}{\left(1-H_{b1}^L\right)} + \frac{\left(\frac{(\beta_2 b)^2\mu_2}{2(\mu_2-\mu_1)}\frac{(\beta_2^2-\sigma^2)b^2\mu_2}{2(\mu_2-\mu_1)}\right)\left(n^2 - H_{b1}^L J_{b1}^T + H_{b1}^L\frac{(\beta_1 b)^2\mu_1}{2(\mu_2-\mu_1)}\right)\left(n^2 - H_{b1}^L + \frac{(\beta_1^2-\sigma^2)b^2\mu_1}{2(\mu_2-\mu_1)}\right)}{\left(1-H_{b1}^L\right)D_2^S}\right]$$

$$C_{22} = \frac{1}{\left(F_{b2}^L - \tilde{H}_{b2}^L\right)\frac{H_n^{(2)}(\alpha_2 b)}{H_n^{(2)}(\alpha_2 a)}}\left[F_{b2}^L + \frac{\left(\frac{(\beta_2^2-\sigma^2)b^2\mu_2}{2(\mu_2-\mu_1)}\right)\left(n^2 - H_{b1}^L J_{b1}^T + H_{b1}^L\frac{(\beta_1 b)^2\mu_1}{2(\mu_2-\mu_1)}\right)}{D_2^S}\right]$$

$$D_2^S = n^2\left(1 - H_{b1}^L\right)\left(1 - J_{b1}^T\right) - \left(n^2 - J_{b1}^T + \frac{(\beta_1 b)^2\mu_1}{2(\mu_2-\mu_1)}\right)\left(n^2 - H_{b1}^L + \frac{(\beta_1^2-\sigma^2)\,b^2\mu_1}{2(\mu_2-\mu_1)}\right)$$

4.3.2 Dynamic stress concentrations

We may now write the explicit expressions for the dynamic stress concentrations due to an incident transverse displacement field. These can then be evaluated to determine the effect of an interface layer. We normalize the dynamic stress concentrations by calculating the incident stress field and computing σ^*. In this case, all stresses will be normalized to the incident shear stress giving:

$$\sigma^*_{rr} = \frac{\left|\sigma^L_{rr} + \sigma^S_{rr}\right|}{\left|\sigma^i_{xy}\right|}$$

$$\sigma^L_{rr} = \sum_{n=0}^{\infty} \frac{2\mu_1 i}{k_1 r^2} \left((k_1 r \sin \phi^s_L) \frac{H^{(1)\prime}_n(k_1 r \sin \phi^s_L)}{H^{(1)}_n(k_1 r \sin \phi^s_L)} - n^2 + \frac{(K_1 r)^2}{2} \right.$$

$$\left. -(k_1 r)^2 \cos^2 \phi^s_L \right) \left(\frac{k_1}{K_1}\right) \frac{J_n(K_1 b \sin \phi^i)}{H^{(1)}_n(k_1 b \sin \phi^s_L)}$$

$$\left(\frac{\Delta^S_1}{\Delta_0}\right) i^n H^{(1)}_n(k_1 r \sin \phi^s_L) e^{in\theta} e^{ik_1 z \cos \phi^s_L}$$

$$\sigma^S_{rr} = \sum_{n=0}^{\infty} -n \frac{2\mu_1}{K_1 r^2} \left(1 - (K_1 r \sin \phi^s_S) \frac{H^{(1)\prime}_n(K_1 r \sin \phi^s_S)}{H^{(1)}_n(K_1 r \sin \phi^s_S)} \right)$$

$$\frac{J_n(K_1 b \sin \phi^i)}{H^{(1)}_n(K_1 b \sin \phi^s_S)} \left(\frac{\Delta^S_2}{\Delta_0}\right) i^n H^{(1)}_n(K_1 r \sin \phi^s_S) e^{in\theta} e^{iK_1 z \cos \phi^s_S} \tag{4.22}$$

The shear stress concentrations can be written as:

$$\sigma^*_{r\theta} = \frac{\left|\sigma^L_{r\theta} + \sigma^S_{r\theta}\right|}{\left|\sigma^i_{xy}\right|}$$

$$\sigma^L_{r\theta} = \sum_{n=0}^{\infty} n \frac{2\mu_1}{k_1 r^2} \left(1 - (k_1 b \sin \phi^s_L) \frac{H^{(1)\prime}_n(k_1 r \sin \phi^s_L)}{H^{(1)}_n(k_1 r \sin \phi^s_L)} \right)$$

$$\left(\frac{k_1}{K_1}\right) \frac{J_n(K_1 b \sin \phi^i)}{H_n^{(1)}(k_1 b \sin \phi_L^s)} \left(\frac{\Delta_1^S}{\Delta_0}\right) i^n H_n^{(1)}(k_1 r \sin \phi_L^s) e^{in\theta} e^{ik_1 z \cos \phi_L^s}$$

$$\sigma_{r\theta}^S = \sum_{n=0}^{\infty} -\frac{2\mu_1 i}{K_1 r^2} \left((K_1 r \sin \phi_S^s) \frac{H_n^{(1)\prime}(K_1 r \sin \phi_S^s)}{H_n^{(1)}(K_1 r \sin \phi_S^s)} - n^2 + \frac{(K_1 r)^2}{2} \cos^2 \phi_S^s \right)$$

$$\frac{J_n(K_1 b \sin \phi^i)}{H_n^{(1)}(K_1 b \sin \phi_S^s)} \left(\frac{\Delta_2^S}{\Delta_0}\right) i^n H_n^{(1)}(K_1 r \sin \phi_S^s) e^{in\theta} e^{iK_1 z \cos \phi_S^s} \tag{4.23}$$

with the hoop stress concentration expressed as:

$$\sigma_{\theta\theta}^* = \frac{\left|\sigma_{\theta\theta}^L + \sigma_{\theta\theta}^S\right|}{\left|\sigma_{xy}^i\right|}$$

$$\sigma_{\theta\theta}^L = \sum_{n=0}^{\infty} -\frac{2\mu_1}{k_1 r^2} \left(\frac{(K_1 r)^2}{2} - (k_1 r)^2 + n^2 - (k_1 r \sin \phi_L^s) \frac{H_n^{(1)\prime}(k_1 r \sin \phi_L^s)}{H_n^{(1)}(k_1 r \sin \phi_L^s)} \right)$$

$$\left(\frac{k_1}{K_1}\right) \frac{J_n(K_1 b \sin \phi^i)}{H_n^{(1)}(k_1 b \sin \phi_L^s)} \left(\frac{\Delta_1^S}{\Delta_0}\right) i^n H_n^{(1)}(k_1 r \sin \phi_L^s) e^{in\theta} e^{ik_1 z \cos \phi_L^s}$$

$$\sigma_{\theta\theta}^S = \sum_{n=0}^{\infty} in \frac{2\mu_1}{K_1 r^2} \left(1 - (K_1 r \sin \phi_S^s) \frac{H_n^{(1)\prime}(K_1 r \sin \phi_S^s)}{H_n^{(1)}(K_1 r \sin \phi_S^s)} \right)$$

$$\frac{J_n(K_1 b \sin \phi^i)}{H_n^{(1)}(K_1 b \sin \phi_S^s)} \left(\frac{\Delta_2^S}{\Delta_0}\right) i^n H_n^{(1)}(K_1 r \sin \phi_S^s) e^{in\theta} e^{iK_1 z \cos \phi_S^s} \tag{4.24}$$

4.4 Incident uncoupled transverse wave

If we have an incident transverse wave of unit amplitude incident upon the layered cylinder, only scattered, refracted, and transmitted displacement fields of the same type will be generated. The incident displacement field is taken as traveling at an angle ϕ^i to the $\hat{z}$ direction and uncoupled to the longitudinal and first transverse field. The incident π_V displacement field, scattered, refracted, and transmitted fields are given by the displacement potentials:

$$\pi_V^i = \sum_{n=0}^{\infty} \epsilon_n i^n J_n(K_1 r \sin\phi^i) e^{in\theta} e^{iK_1 z \cos\phi^i}$$

$$\pi_V^s = \sum_{n=0}^{\infty} C_{1n} i^n H_n^{(1)}(K_1 r \sin\phi_V^s) e^{in\theta} e^{iK_1 z \cos\phi_V^s}$$

$$\pi_V^r = \sum_{n=0}^{\infty} C_{2n} i^n H_n^{(1)}(K_2 r \sin\phi_V^r) e^{in\theta} e^{iK_2 z \cos\phi_V^r}$$

$$\tilde{\pi}_V^r = \sum_{n=0}^{\infty} \tilde{C}_{2n} i^n H_n^{(2)}(K_2 r \sin\phi_V^r) e^{in\theta} e^{iK_2 z \cos\phi_V^r}$$

$$\pi_V^t = \sum_{n=0}^{\infty} C_{3n} i^n J_n(K_3 r \sin\phi_V^t) e^{in\theta} e^{iK_3 z \cos\phi_V^t} \tag{4.25}$$

By applying the boundary conditions, we arrive at the four equations in four unknowns to solve. In matrix form we can write them as $Nx = c$. Filling the matrix N with the expressions for the displacements and normal surface tractions at the two interfaces gives the following values for the elements of the matrix N:

$$N[1,1] = (K_1 b \sin\phi_V^s)^2$$

$$N[1,2] = -(K_2 b \sin\phi_V^r)^2 \frac{H_n^{(1)}(K_2 b \sin\phi_V^r)}{H_n^{(1)}(K_2 a \sin\phi_V^r)}$$

$$N[1,3] = -(K_2 b \sin\phi_V^r)^2 \frac{H_n^{(2)}(K_2 b \sin\phi_V^r)}{H_n^{(2)}(K_2 a \sin\phi_V^r)}$$

$$N[1,4] = 0$$

$$N[2,1] = \frac{2\mu_1}{b}(K_1 b \sin\phi_V^s) \frac{H_n^{(1)\prime}(K_1 b \sin\phi_V^s)}{H_n^{(1)}(K_1 b \sin\phi_V^s)} \frac{(K_1 b)^2}{2} \left(1 - 2\cos^2\phi_V^s\right)$$

$$N[2,2] = -\frac{2\mu_2}{b}(K_2 b\sin\phi_V^r)\frac{H_n^{(1)\prime}(K_2 b\sin\phi_V^r)}{H_n^{(1)}(K_2 b\sin\phi_V^r)}\frac{(K_2 b)^2}{2}\left(1-2\cos^2\phi_V^r\right)$$

$$\frac{H_n^{(1)}(K_2 b\sin\phi_V^r)}{H_n^{(1)}(K_2 a\sin\phi_V^r)}$$

$$N[2,3] = -\frac{2\mu_2}{b}(K_2 b\sin\phi_V^r)\frac{H_n^{(2)\prime}(K_2 b\sin\phi_V^r)}{H_n^{(2)}(K_2 b\sin\phi_V^r)}\frac{(K_2 b)^2}{2}\left(1-2\cos^2\phi_V^r\right)$$

$$\frac{H_n^{(2)}(K_2 b\sin\phi_V^r)}{H_n^{(2)}(K_2 a\sin\phi_V^r)}$$

$$N[2,4] = 0 \qquad N[3,1] = 0$$

$$N[3,2] = -\frac{2\mu_2}{a}(K_2 a\sin\phi_V^r)\frac{H_n^{(1)\prime}(K_2 a\sin\phi_V^r)}{H_n^{(1)}(K_2 a\sin\phi_V^r)}\frac{(K_2 a)^2}{2}\left(1-2\cos^2\phi_V^r\right)$$

$$N[3,3] = -\frac{2\mu_2}{a}(K_2 a\sin\phi_V^r)\frac{H_n^{(2)\prime}(K_2 a\sin\phi_V^r)}{H_n^{(2)}(K_2 a\sin\phi_V^r)}\frac{(K_2 a)^2}{2}\left(1-2\cos^2\phi_V^r\right)$$

$$N[3,4] = \frac{2\mu_3}{a}\left(K_3 a\sin\phi_V^t\right)\frac{J_n'(K_3 a\sin\phi_V^t)}{J_n(K_3 a\sin\phi_V^t)}\frac{(K_3 a)^2}{2}\left(1-2\cos^2\phi_V^t\right)$$

$$N[4,1] = 0$$

$$N[4,2] = N[4,3] = (K_2 a\sin\phi_V^r)^2$$

$$N[4,4] = \left(K_3 a\sin\phi_V^t\right)^2 \tag{4.26}$$

In the above matrix, we have factored the radial functions as before in the scattered, refracted, and transmitted fields. We can now express the incident field in terms of the boundary conditions. The c column vector for an incident π_V transverse wave has the following components after factoring out the radial function common to all terms:

$$\begin{aligned}
c[1] &= \left(K_1 b \sin\phi^i\right)^2 \\
c[2] &= \frac{2\mu_1}{b}\left(K_1 b \sin\phi^i\right)\frac{J_n'(K_1 b \sin\phi^i)}{J_n(K_1 b \sin\phi^i)}\frac{(K_1 b)^2}{2}\left(1 - 2\cos^2\phi^i\right) \\
c[3] &= c[4] = 0
\end{aligned} \tag{4.27}$$

and modal coefficients of the potentials can then be expressed in terms of the determinants as:

$$\begin{aligned}
C_1 &= -\epsilon_n \frac{J_n(K_1 b \sin\phi^i)}{H_n^{(1)}(K_1 b \sin\phi_V^s)}\left(\frac{\Delta_1^V}{\Delta_0}\right) \\
C_2 &= \epsilon_n \frac{J_n(K_1 b \sin\phi^i)}{H_n^{(1)}(K_2 b \sin\phi_V^r)}\left(\frac{K_2}{K_1}\right)^2\left(\frac{\Delta_2^V}{\Delta_0}\right) \\
\tilde{C}_2 &= \epsilon_n \frac{J_n(K_1 b \sin\phi^i)}{H_n^{(2)}(K_2 b \sin\phi_V^r)}\left(\frac{K_2}{K_1}\right)^2\left(\frac{\tilde{\Delta}_2^V}{\Delta_0}\right) \\
C_3 &= -\epsilon_n \frac{J_n(K_1 b \sin\phi^i)}{J_n(K_3 a \sin\phi_V^t)}\left(\frac{K_3}{K_1}\right)^2\left(\frac{\Delta_3^V}{\Delta_0}\right)
\end{aligned} \tag{4.28}$$

4.4.1 Scattered field and stress concentration

The scattered field from the cylindrical interface will be of the same type as incident in this case. The scattered displacement potential is given below after substitution of the expression for the modal coefficients as

$$\pi_V^s = \sum_{n=0}^{\infty} -\epsilon_n \frac{J_n(K_1 b \sin\phi^i)}{H_n^{(1)}(K_1 b \sin\phi_V^s)}\left(\frac{\Delta_1^V}{\Delta_0}\right) i^n H_n^{(1)}(K_1 r \sin\phi_V^s) e^{in\theta} e^{iK_1 z \cos\phi_V^s} \tag{4.29}$$

with the expressions for Δ_0,and Δ_1^V given as:

$$\begin{aligned}
\Delta_1^V \;=\; & \left(F_2(K_1 b\sin\phi_{1v})^2\frac{(K_2 b)^2}{2}\left(1-2\cos^2\phi_{2v}\right)-J_n(\beta_1 b)\frac{\mu_1}{\mu_2}\frac{(K_1 b)^2}{2}\left(1-2\cos^2\phi_{1v}\right)(K_2 b\sin\phi_{2v})^2\right)\\
& \left(H_n^{(2)}(\beta_2 a)\frac{(K_2 a)^2}{2}\left(1-2\cos^2\phi_{2v}\right)(K_3 a\sin\phi_{3v})^2-J_n(\beta_3 a)\frac{\mu_3}{\mu_2}(K_2 a\sin\phi_{2v})^2\frac{(K_3 a)^2}{2}\left(1-2\cos^2\phi_{3v}\right)\right)\\
& +F_1\frac{H_n^{(2)}(\beta_2 b)}{H_n^{(2)}(\beta_2 a)}\frac{(K_2 a)^2}{2}\left(1-2\cos^2\phi_{2v}\right)(K_3 a\sin\phi_{3v})^2\\
& \left(J_n(\beta_1 b)\frac{\mu_1}{\mu_2}\frac{(K_1 b)^2}{2}\left(1-2\cos^2\phi_{1v}\right)(K_2 b\sin\phi_{2v})^2-H_n^{(2)}(\beta_2 b)(K_1 b\sin\phi_{1v})^2\frac{(K_2 b)^2}{2}\left(1-2\cos^2\phi_{2v}\right)\right) \qquad (4.29)
\end{aligned}$$

$$\begin{aligned}
\Delta_0 \;=\; & \left(F_2(K_1 b\sin\phi_{1v})^2\frac{(K_2 b)^2}{2}\left(1-2\cos^2\phi_{2v}\right)-H_n^{(1)}(\beta_1 b)\frac{\mu_1}{\mu_2}\frac{(K_1 b)^2}{2}\left(1-2\cos^2\phi_{1v}\right)(K_2 b\sin\phi_{2v})^2\right)\\
& \left(H_n^{(2)}(\beta_2 a)\frac{(K_2 a)^2}{2}\left(1-2\cos^2\phi_{2v}\right)(K_3 a\sin\phi_{3v})^2-J_n(\beta_3 a)\frac{\mu_3}{\mu_2}(K_2 a\sin\phi_{2v})^2\frac{(K_3 a)^2}{2}\left(1-2\cos^2\phi_{3v}\right)\right)\\
& +F_1\frac{H_n^{(2)}(\beta_2 b)}{H_n^{(2)}(\beta_2 a)}\frac{(K_2 a)^2}{2}\left(1-2\cos^2\phi_{2v}\right)(K_3 a\sin\phi_{3v})^2\\
& \left(H_n^{(1)}(\beta_1 b)\frac{\mu_1}{\mu_2}\frac{(K_1 b)^2}{2}\left(1-2\cos^2\phi_{1v}\right)(K_2 b\sin\phi_{2v})^2-H_n^{(2)}(\beta_2 b)(K_1 b\sin\phi_{1v})^2\frac{(K_2 b)^2}{2}\left(1-2\cos^2\phi_{2v}\right)\right) \qquad (4.30)
\end{aligned}$$

$$F_1=(\beta_2 a)\left[\frac{H_n^{(1)\prime}(\beta_2 a)H_n^{(2)}(\beta_2 a)-H_n^{(2)\prime}(\beta_2 a)H_n^{(1)}(\beta_2 a)}{H_n^{(1)}(\beta_2 b)H_n^{(2)}(\beta_2 a)-H_n^{(2)}(\beta_2 b)H_n^{(1)}(\beta_2 a)}\right] \qquad F_2=(\beta_2 b)\left[\frac{H_n^{(1)\prime}(\beta_2 b)H_n^{(2)}(\beta_2 a)-H_n^{(2)\prime}(\beta_2 b)H_n^{(1)}(\beta_2 a)}{H_n^{(1)}(\beta_2 b)H_n^{(2)}(\beta_2 a)-H_n^{(2)}(\beta_2 b)H_n^{(1)}(\beta_2 a)}\right]$$

$$\beta_{1,2,3}=K_{1,2,3}\sin\phi_{1v,2v,3v}$$

The expression for the modal coefficients of the scattered field displacement potential can now be used to explicitly write the expression for the dynamic stress concentration as:

$$\sigma^*_{rz} = \frac{|\sigma^V_{rz}|}{|\sigma^i_{xz}|}$$

where

$$\sigma^S_{rz} = \sum_{n=0}^{\infty} -\left(\frac{i}{K_1 r}\right)^2 \frac{2\mu_1}{r}\left(K_1 r \sin\phi^s_V\right)\frac{(K_1 r)^2}{2}\left(1 - 2\cos^2\phi^s_V\right)$$

$$\epsilon_n \frac{J_n(K_1 b \sin\phi^i)}{H^{(1)}_n(K_1 b \sin\phi^s_V)}\left(\frac{\Delta^V_1}{\Delta_0}\right) i^n H^{(1)\prime}_n(K_1 r \sin\phi^s_V) e^{in\theta} e^{iK_1 z \cos\phi^s_V} \tag{4.32}$$

4.5 Applications to fiber composites

There are many applications of this analysis of dynamical stress concentrations to fiber composite materials. The three that have been noted previously: boron fibers in an epoxy matrix, silicon-carbide fibers in a metal matrix, and tungsten fibers in steel, are the cases that will be studied. In the first case, the boron fiber is the shell material, a tungsten fiber forms the core, and the exterior is an epoxy resin. The second case will be a silicon-carbide shell, a carbon core, and an aluminum exterior Last we will look at a tungsten fiber embedded in steel with a thin reaction layer between the tungsten and the steel. In the first case, a boron fiber in epoxy, the dependence of radial, shear, and hoop stresses are given in figures 4.2-4.4. In each case, the dynamic stress concentrations are given as a function of $k_1 b$ and are shown for two angles, $\theta = \pi$ and $\theta = \frac{\pi}{2}$. These are the directions facing back toward the incident field and facing perpendicular to the incident field. The dynamic stress concentrations are normalized to the incident stress field based on a unit amplitude incident displacement wave of the longitudinal type. The normal dynamic stress concentrations show some definite resonant peaks and nulls. The configuration seems to have a radial stress peak at $k_1 b = 6.6$ in the backward direction and null in the perpendicular direction. This means that the incident displacement field is reflected back towards the source at this value of $k_1 b$.

In the second case, a silicon carbide fiber in aluminum, the dependence of radial, shear, and shear stresses are given in figures 4.5-4.7. In each case, the dynamic stress concentrations are given as a function of $k_1 b$ and are shown for two angles, $\theta = \pi$ and $\theta = \frac{\pi}{2}$ as in the first case. Here, a resonant peak occurs in the perpendicular direction to the incident displacement field direction and at an extremely high value of $k_1 b$ near nine. This configuration seems to preferentially intensify stress levels in the perpendicular direction.

Lastly, a tungsten fiber in steel is studied for the dependence of radial, shear, and hoop stresses as given in figures 4.8-4.10. Perhaps the most interesting of the three, this interface configuration gives rise to complex and strong $k_1 b$ dependance of the scattered stress levels. The existence of the greater number of resonances and nulls is not due to the thin interface layer seperating the tungsten from the steel, but from the combination of these two specific materials. Dynamic stress concentrations are shown in figures 4.11-4.13 for the tungsten fiber in steel without an interface layer. In general, the scattered stress levels are all higher for the unlayered fiber throughout the frequency range. The interface seems to broaden the nulls and reduce the peaks of the dynamic stress concentrations in this configuration.

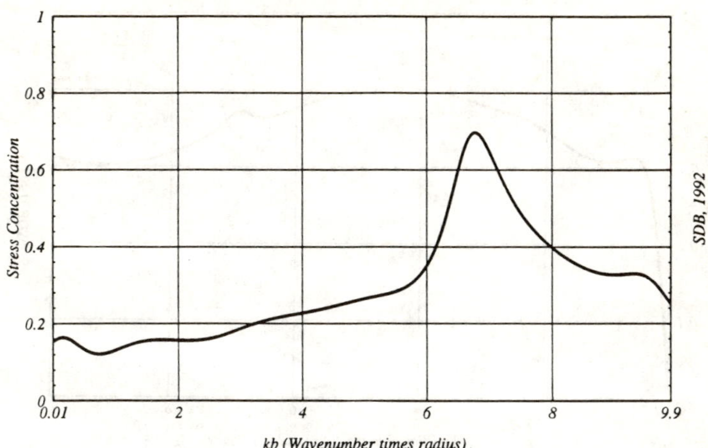

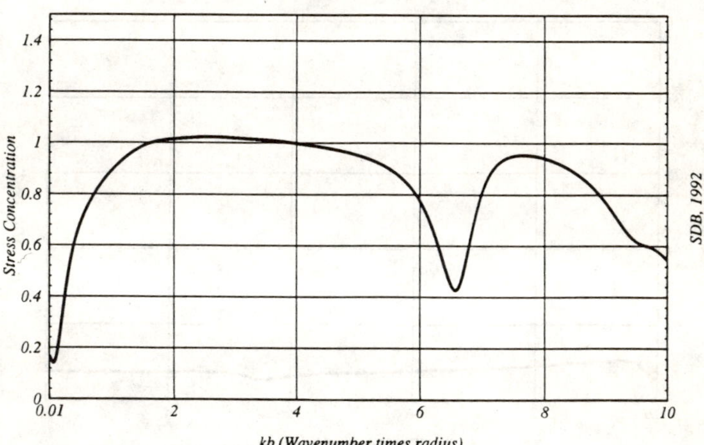

Figure 4.2: Dynamic stress concentrations: boron-radial 90deg and 180degs.

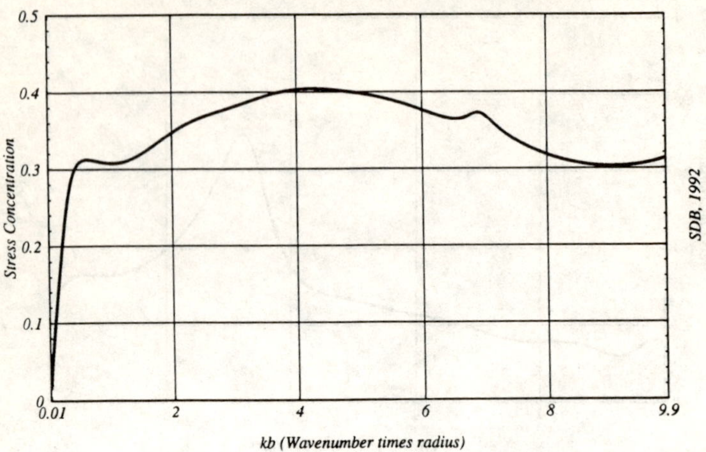

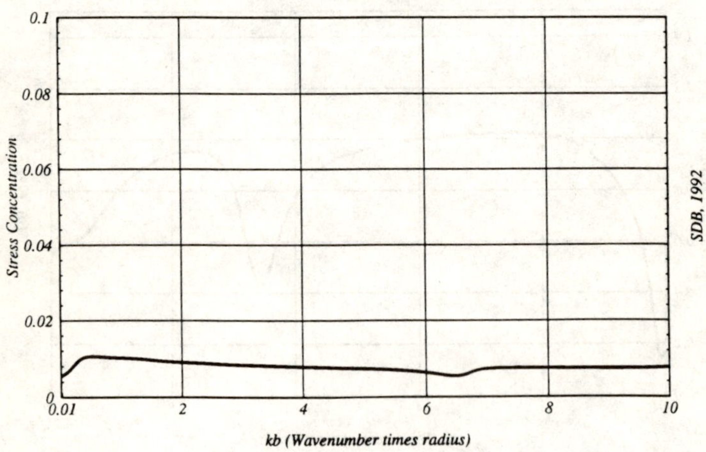

Figure 4.3: Dynamic stress concentrations: boron-shear 90deg and 180degs.

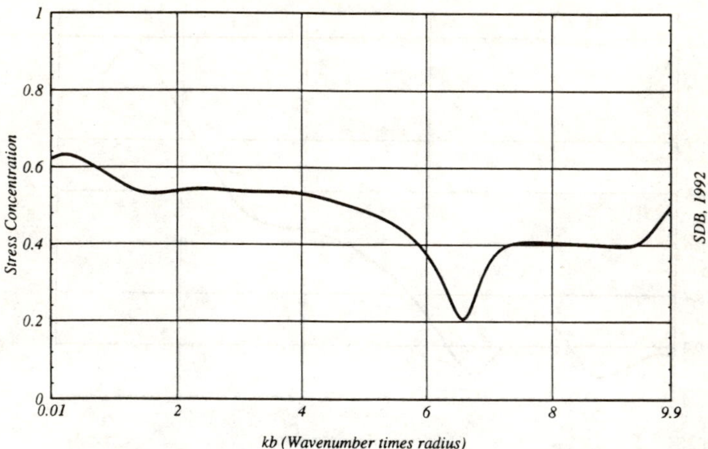

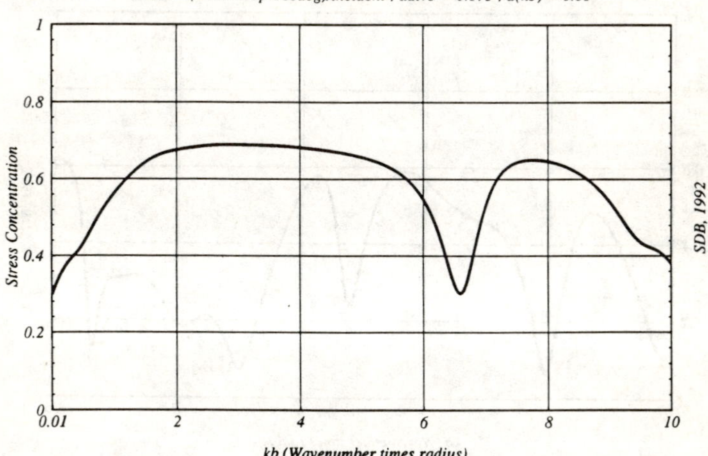

Figure 4.4: Dynamic stress concentrations: boron-hoop 90deg and 180degs.

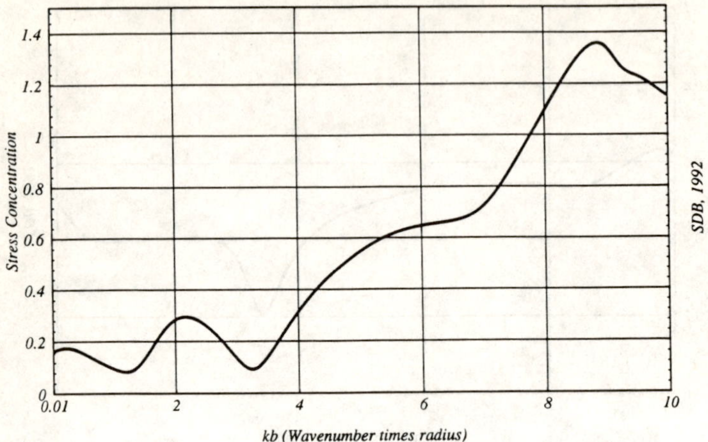

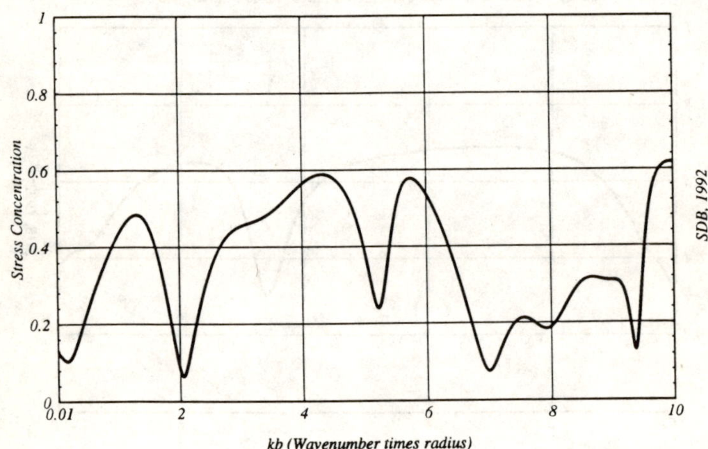

Figure 4.5: Dynamic stress concentrations: SiC-radial 90deg and 180degs.

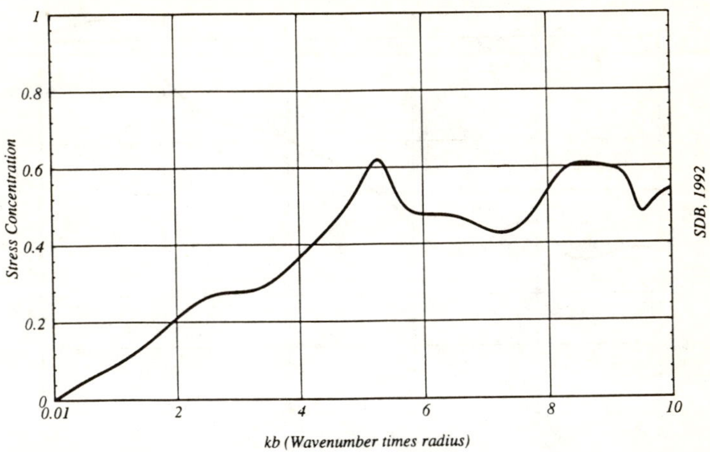

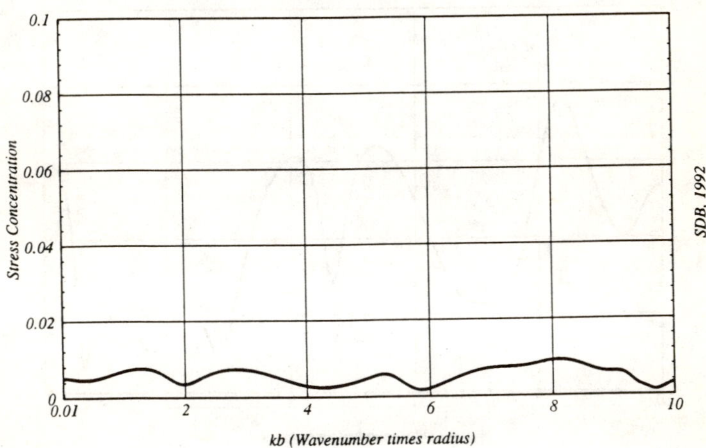

Figure 4.6: Dynamic stress concentrations: SiC-shear 90deg and 180degs.

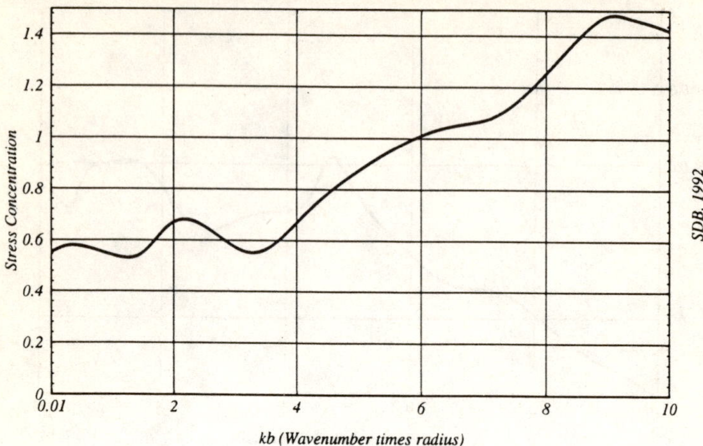

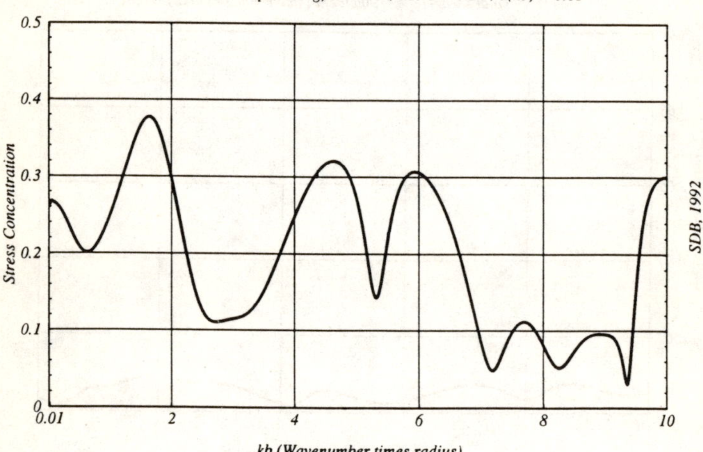

Figure 4.7: Dynamic stress concentrations: SiC-hoop 90deg and 180degs.

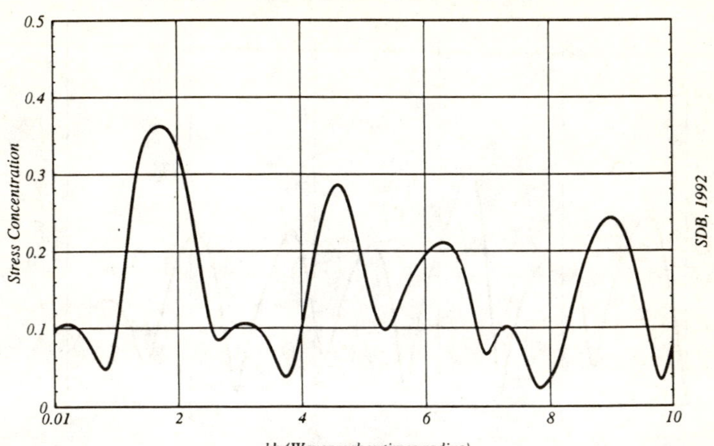

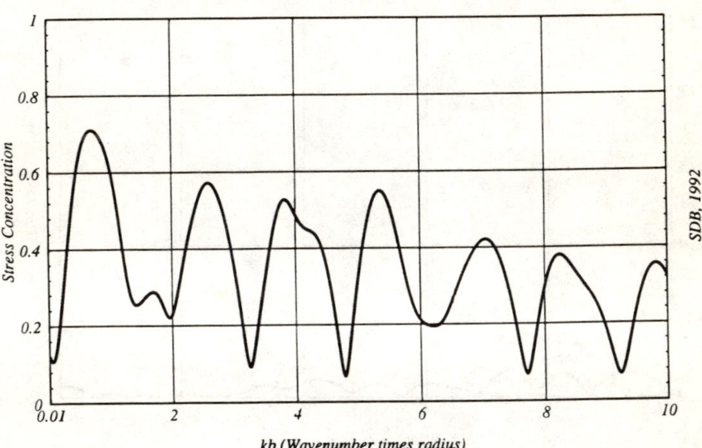

Figure 4.8: Dynamic stress concentrations: tungsten-radial 90deg and 180degs.

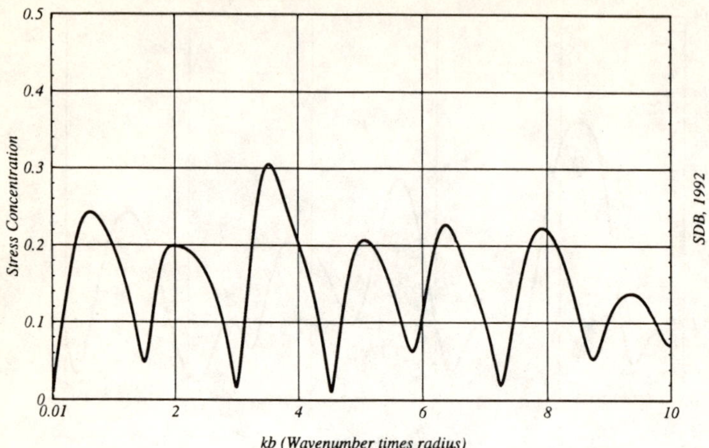

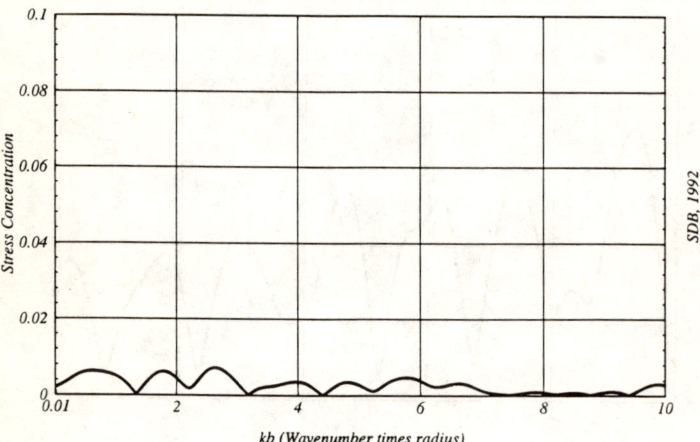

Figure 4.9: Dynamic stress concentrations: tungsten-shear 90deg and 180degs.

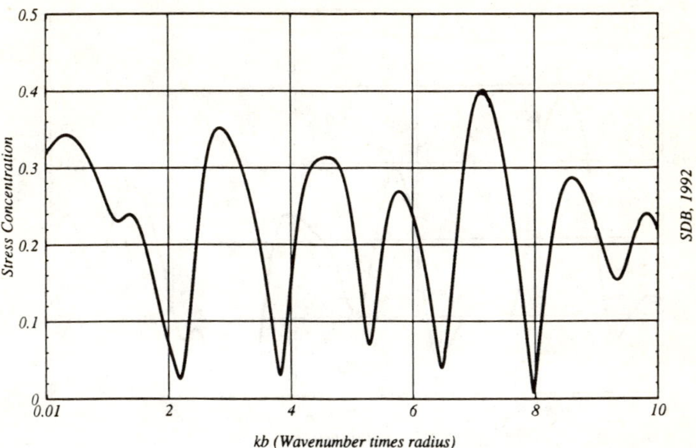

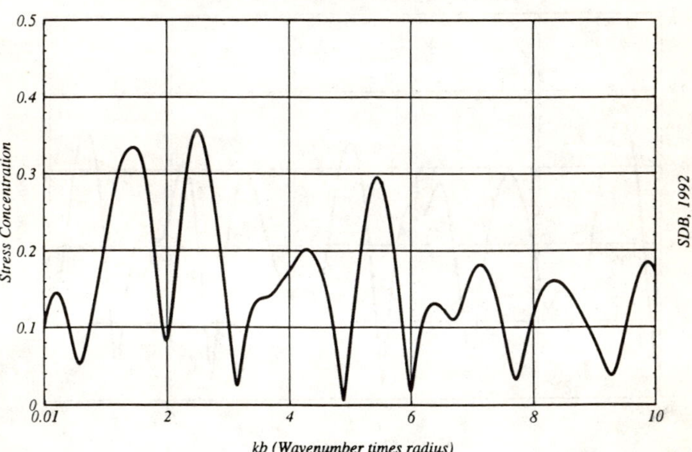

Figure 4.10: Dynamic stress concentrations: tungsten-hoop 90deg and 180degs.

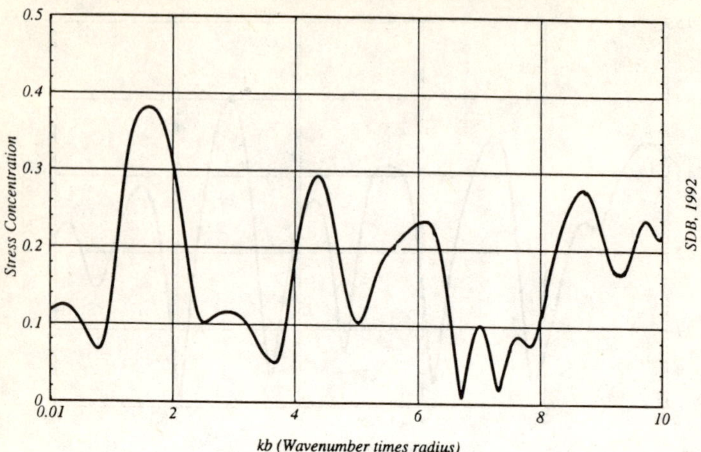

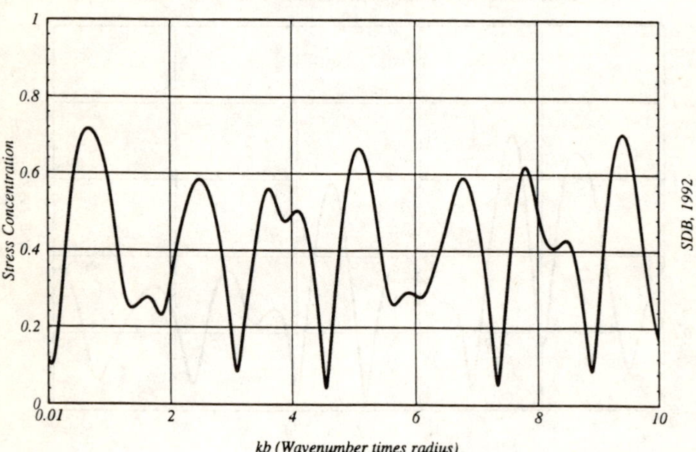

Figure 4.11: Dyn. stress conc.: tungsten-no interface-radial 90deg and 180degs.

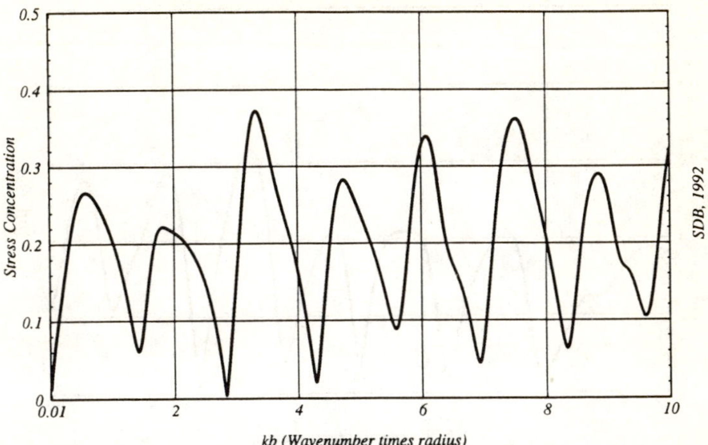

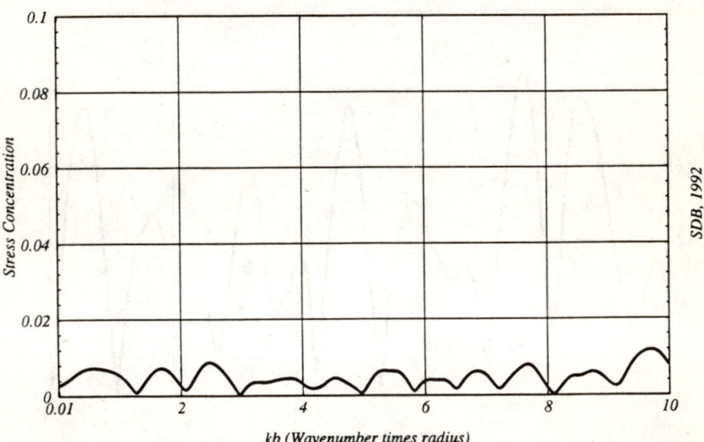

Figure 4.12: Dyn. stress conc.: tungsten-no interface-shear 90deg and 180degs.

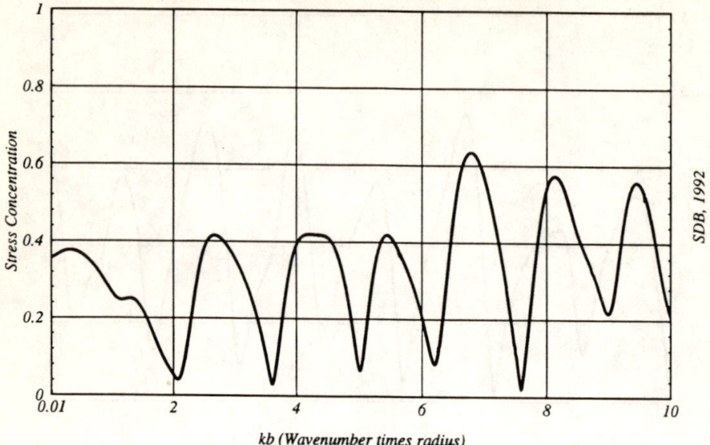

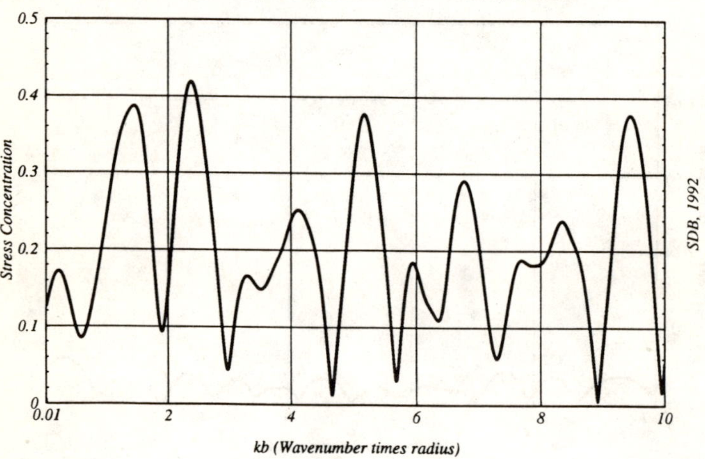

Figure 4.13: Dyn. stress conc.: tungsten-no interface-hoop 90deg and 180degs.

4.6 Special cases and verification

There are a few special cases of interest in checking the results of this analysis. One particularly interesting is to let the core material have a density much less than the shell or exterior mediums. This gives an approximation to the hollow cylinder, and brings the determinants of boundary conditions to a near zero value testing the extremes of applicability of this analysis. In the first case, we will use a hollow cylinder with a thin steel shell embedded in a concrete exterior. The scattered stress fields will be studied in the concrete on the steel interface.

Since a hollow cylinder can have no radial stress components on the inside surface, we expect an incident longitudinal displacement field to be perfectly reflected back in the direction of the incident field independant of the frequency. If we look at figure 4.14, we see that the scattered radial dynamic stress concentration in the cement, on the interface with the steel just outside the free surface, is near unity for k_1b greater than five and at very small k_1b jumps to 3/4. It appears that the scattered radial stress field is not unity over a large k_1b range that allows for the shell to scatter the incident field in different directions. The hoop stress (see figure 4.16) is greater at 90 degrees than 180degress for small k_1b. This can be interpreted as a near static loading. As in the static tension load on a hollow cylinder, the largest hoop stress is on the hollow cylinder at 90 degrees.

As a final comparison, we take the analysis to an even further extreme by taking a thin ($a/b = 0.99$) shell of aluminum surrounding a cavity in an aluminum exterior and evaluate near the free surface. The radial scattered stress due to an incident longitudinal displacement field (figure 4.17) is unity for all but a small range of k_1b in the incident field direction cancelling the incident field and giving a net zero radial stress as required on the free surface. The hoop stress concentrations are given in figure 4.18 and corresponds well with the results of Y.H. Pao and C.C. Mow for the hollow cylinder [83].

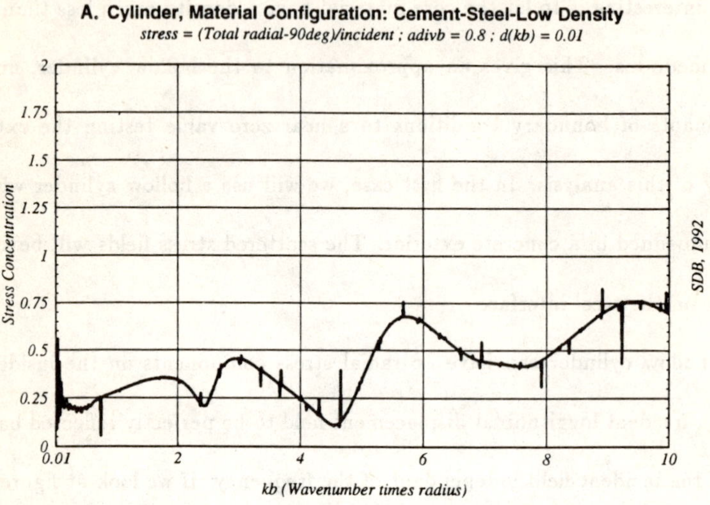

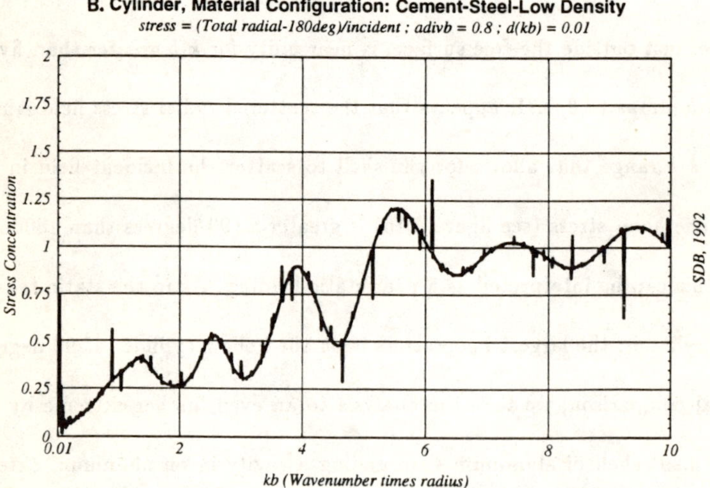

Figure 4.14: Limiting case: radial dyn. stress conc. cement-steel-hollow.

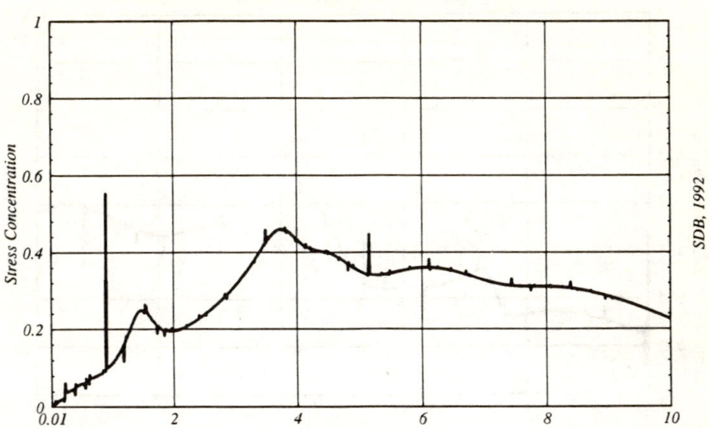

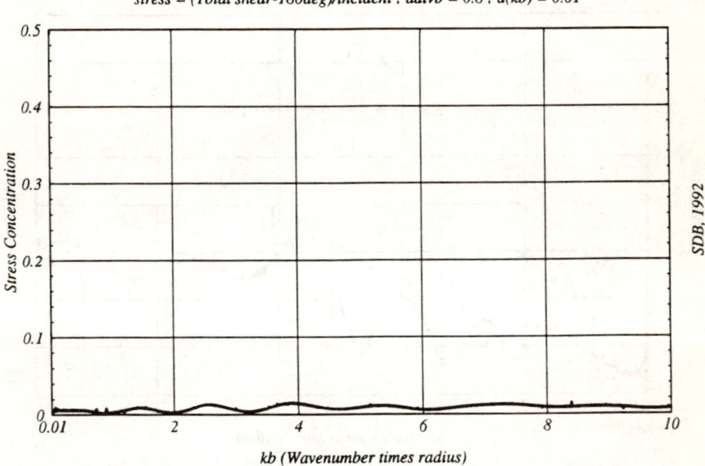

Figure 4.15: Limiting case: shear dyn. stress conc. cement-steel-hollow.

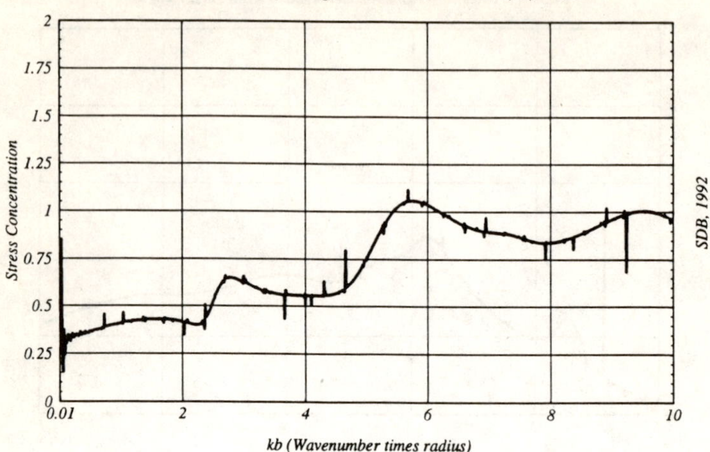

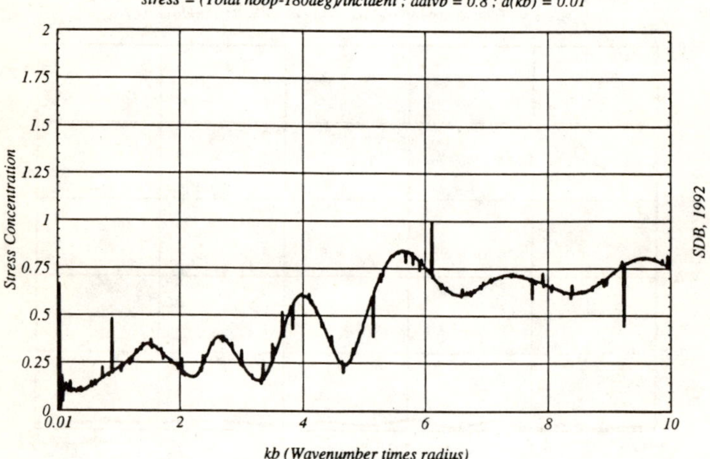

Figure 4.16: Limiting case: hoop dyn. stress conc. cement-steel-hollow.

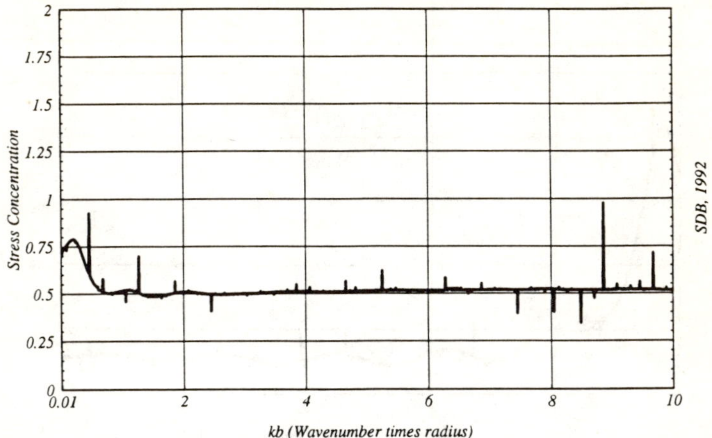

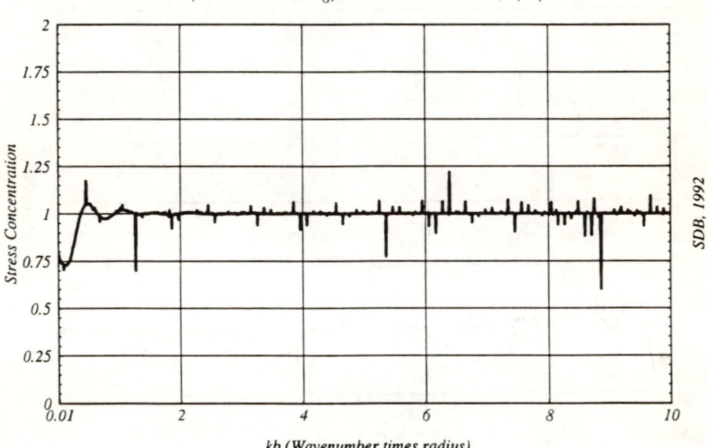

Figure 4.17: Extreme limiting case: radial dyn. stress conc. Al-Al-hollow.

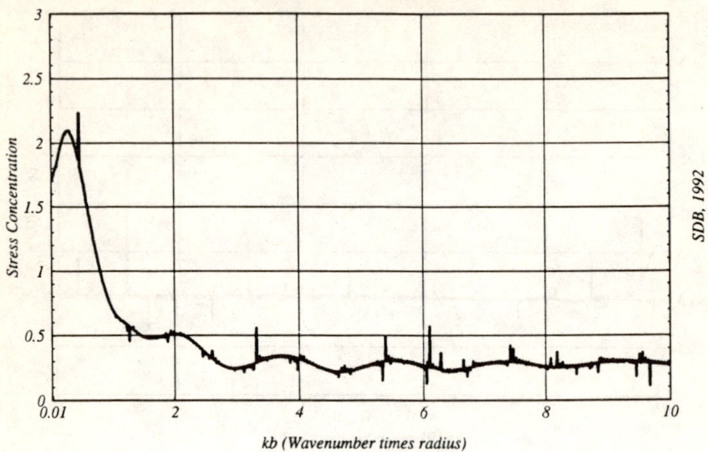

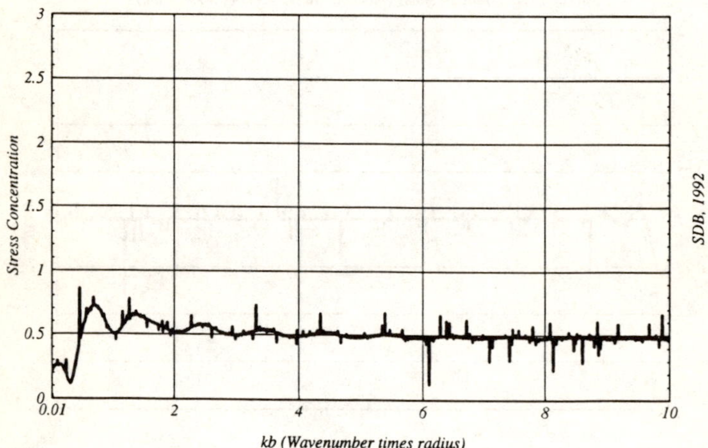

Figure 4.18: Extreme limiting case: hoop dyn. stress conc. Al-Al-hollow.

Chapter 5

LAYERED SPHERICAL INTERFACE

5.1 Introduction

In the analysis of scattering of elastic waves from a spherical layered interface, we are interested in modeling particle reinforced composites and the interface structures that develop as a result of processing or are intentionally introduced between the particle and matrix [55],[56]. Perhaps the best known example of this is in cements. A water paste surrounds the particles of stone, bonding them to the matrix [85]. The study of particle reinforced composites is the primary area of interest in the present analysis, and will be the focus. The problem geometry is seen in figure 5.1.

In this section, we will review the formal solution to the scattering of elastic waves from a layered spherical interface and provide the analytic expressions for the displacement potentials. Three different displacement fields will be incident on the spherical interface layer: a longitudinal displacement field, a transverse displacement field that couples to the longitudinal, and an uncoupled transverse displacement field. In this case, we have taken $\vec{a} = r\hat{e}_r$ in calculating the curls for the two transverse displacement potentials.

To solve the scattering problem, we will need to set up the boundary conditions formulated in chapter 2 for the spherical interface. The displacements and normal surface tractions are the appropriate field variables that must be continuous across these spherical boundaries. The boundary conditions involve derivatives of spherical and Ricatti Bessel and Hankel functions, and in order to simplify the expressions, we redefine these deriva-

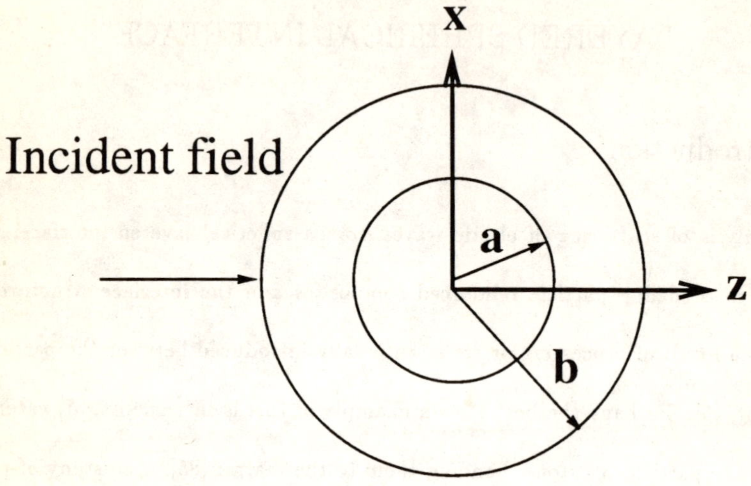

Figure 5.1: The problem geometry for spherical interfaces.

tive by forming a log-like ratio of the derivative times argument over the function. For example, the Ricatti-Bessel function ratio $k_1 b \frac{\psi_l'(k_1 b)}{\psi_l(k_1 b)}$ is rewritten as ψ_{b1}^L. Below we will use this notation, but not substitute in the appropriate radial functions quite yet. An intermediate symbolic step is necessary to write the expression for the displacements and normal surface tractions prior to evaluation at the boundaries. Rather than ψ we will use the symbol $z_l^{L,S,V}$ to represent the log-like derivative of the radial functions, identifying the type of field by a superscript L, S, V. While this may seem a tedious first step to take, the writing of the displacements and normal surface tractions in this form, then evaluating at the appropriate boundaries, eliminates a possible source of error in the manipulation of these lengthy expressions.

To restate the definitions of some of the terms in the expressions, the subscript l refers to the summation index of the series solution to the Helmholtz equation. The displacement and stress fields will be an infinite series with a sum over the index l. The longitudinal displacement potential is written as π_L, coupled transverse as π_V, and uncoupled transverse as π_S.

$$
\begin{aligned}
\sigma_{rr} &= 2\mu\left[\left(\frac{i}{kr^2}\right)\left(\frac{(Kr)^2}{2} - (l(l+1)+2) + 2z_l^L\right)(r\pi_L) + \quad 0 \quad + \left(\frac{-i}{Kr}\right)^2 (-l(l+1))\left(2 - z_l^V\right)(r\pi_V)\right] \qquad (5.1)\\
\sigma_{r\theta} &= 2\mu\left[\left(\frac{i}{kr^2}\right)\left(2 - z_l^L\right)\partial_\theta(r\pi_L) + \left(\frac{-i}{Kr^2\sin\theta}\right)\left(1 - 2z_l^S\right)\partial_\phi(r\pi_S) + \left(\frac{-i}{Kr}\right)^2\left(l(l+1) - \frac{(Kr)^2}{2} - z_l^V\right)\partial_\theta(r\pi_V)\right]\\
\sigma_{r\phi} &= 2\mu\left[\left(\frac{i}{kr^2\sin\theta}\right)\left(2 - z_l^L\right)\partial_\phi(r\pi_L) + \left(\frac{-i}{Kr^2}\right)\left(1 - \frac{z_l^S}{2}\right)\partial_\theta(r\pi_S) + \left(\frac{-i}{Kr}\right)^2\frac{1}{\sin\theta}\left(l(l+1) - \frac{(Kr)^2}{2} - z_l^V\right)\partial_\phi(r\pi_V)\right]
\end{aligned}
$$

$$
\begin{aligned}
u_r &= \left(\frac{1}{kr}\right)^2\left(1 - z_l^L\right)(r\pi_L) + \quad 0 \quad + \left(\frac{1}{Kr}\right)^2 (l(l+1))(r\pi_V)\\
u_\theta &= \left(\frac{-1}{(kr)^2}\right)(\partial_\theta)(r\pi_L) + \left(\frac{1}{Kr\sin\theta}\right)(\partial_\phi)(r\pi_S) + \left(\frac{-i}{Kr}\right)^2 (iKr\cos\phi)\,\partial\theta(r\pi_V)\\
u_\phi &= \left(\frac{-i}{kr}\right)(ikr\cos\phi)(r\pi_L) + \quad 0 \quad + \left(\frac{-i}{Kr}\right)^2 (Kr\sin\phi)^2 (r\pi_V) \qquad (5.2)
\end{aligned}
$$

$$
\begin{aligned}
z_l^L &= \frac{(kr)\psi_l'(kr)}{\psi_l(kr)} \quad or \quad \frac{(kr)\zeta_l^{(1,2)\prime}(kr)}{\zeta^{(1,2)}z_l(kr)}\\
z_l^S &= \frac{(Kr)\psi_l'(Kr)}{\psi_l(Kr)} \quad or \quad \frac{(Kr)\zeta_l^{(1,2)\prime}(Kr)}{\zeta_l^{(1,2)}(Kr)}\\
z_l^V &= \frac{(Kr)\psi_l'(Kr)}{\psi_l(Kr)} \quad or \quad \frac{(Kr)\zeta_l^{(1,2)\prime}(Kr)}{\zeta_l^{(1,2)}(Kr)} \qquad (5.3)
\end{aligned}
$$

The bounding surfaces are constant radial surfaces located at $r = a$ for the inner boundary and $r = b$ for the outer boundary. Incident displacement fields are taken as propagating in the positive $\hat{z}$ direction. The proper boundary conditions for each interface, as discussed in chapter 2, are the three displacements and the radial components of stress given above. This gives a maximum of six boundary conditions per interface. The π_S transverse displacement wave is fully decoupled from the longitudinal displacement π_L and transverse displacement π_V waves on a spherical boundary, reducing the problem to a system of four equations for the π_L and π_V displacement potentials and a system of two equations for the π_S displacement potentials at each interface. The two interface problem will therefore lead to a system of eight equations for displacements and normal surface tractions with eight unknown coefficients of the π_L and π_V displacement potentials, and a system of four equations and four unknown modal coefficients for the π_S displacement potentials.

We can write these systems in symbolic form as $Mx = b$, for the incident longitudinal and coupled transverse displacement fields and $Nx = c$ for the incident second transverse displacement field. The algebraic solutions will be written in terms of the ratio of the determinant of the boundary condition matrix, and the substitution matrices via Cramer's rule.

5.2 Incident longitudinal wave

If we have an incident longitudinal (compressional-ultrasonic) displacement wave of unit amplitude traveling in the $\hat{z}$ direction, scattered, refracted, and transmitted displacement fields will be generated. We begin by writing the incident planar displacement wave as the displacement potential

$$(k_1r)\pi_L^i = \sum_{l=0}^{\infty} i^{l+1}\,(2l+1)\,\psi_l(k_1r)P_l(\cos\theta) \tag{5.4}$$

The eight unknown modal coefficients of the displacement fields are given by the displacement potentials defined in chapter 2. We repeat them here for convenience, noting that we have only the modal coefficients due to an incident longitudinal displacement field to solve for. The potentials are given by:

$$\begin{aligned}
(k_1r)\pi_L^s &= \sum_{l=0}^{\infty} a_{1l} i^{l+1} \zeta_l^{(1)}(k_1r)P_l(\cos\theta) \\
(K_1r)\pi_V^s &= \sum_{l=1}^{\infty} c_{1l} i^{l-1} \zeta_l^{(1)}(K_1r)P_l^{(1)}(\cos\theta)\cos\phi \\
(k_2r)\pi_L^r &= \sum_{l=0}^{\infty} a_{2l} i^{l+1} \zeta_l^{(1)}(k_2r)P_l(\cos\theta) \\
(K_2r)\pi_V^r &= \sum_{l=1}^{\infty} c_{2l} i^{l-1} \zeta_l^{(1)}(K_2r)P_l^{(1)}(\cos\theta)\cos\phi \\
(k_2r)\tilde{\pi}_L^r &= \sum_{l=0}^{\infty} \tilde{a}_{2l} i^{l+1} \zeta_l^{(2)}(k_2r)P_l(\cos\theta) \\
(K_2r)\tilde{\pi}_V^r &= \sum_{l=1}^{\infty} \tilde{c}_{2l} i^{l-1} \zeta_l^{(2)}(K_2r)P_l^{(1)}(\cos\theta)\cos\phi \\
(k_3r)\pi_L^t &= \sum_{l=0}^{\infty} a_{3l} i^{l+1} \psi_l(k_3r)P_l(\cos\theta) \\
(K_3r)\pi_V^t &= \sum_{l=1}^{\infty} c_{3l} i^{l-1} \psi_l(K_3r)P_l^{(1)}(\cos\theta)\cos\phi
\end{aligned} \tag{5.5}$$

By applying the boundary conditions, we arrive at the eight equations for displacement and normal surface tractions, and eight unknown modal coefficients to solve for. Filling the matrix M with the boundary conditions of the displacement fields, we write the most general form for bounding surfaces at $r = a$ and $r = b$. The radial functions in the interior and exterior regions are evaluated at the boundaries and factored out of the boundary condition matrix, as are the radial functions for the shell region evaluated at the inner boundary. The components of the matrix M are:

$$M[1,1] = \frac{1}{(k_1 b)^2}$$

$$M[1,2] = -\frac{1}{(K_1 b)^2}\left((K_1 b)\frac{\zeta_l^{(1)\prime}(K_1 b)}{\zeta_l^{(1)}(K_1 b)}\right)$$

$$M[1,3] = -\frac{1}{(k_2 b)^2}\left(\frac{\zeta_l^{(1)}(k_2 b)}{\zeta_l^{(1)}(k_2 a)}\right)$$

$$M[1,4] = -\frac{1}{(k_2 b)^2}\left(\frac{\zeta_l^{(2)}(k_2 b)}{\zeta_l^{(2)}(k_2 a)}\right)$$

$$M[1,5] = \frac{1}{(K_2 b)^2}\left((K_2 b)\frac{\zeta_l^{(1)\prime}(K_2 b)}{\zeta_l^{(1)}(K_2 b)}\frac{\zeta_l^{(1)}(K_2 b)}{\zeta_l^{(1)}(K_2 a)}\right)$$

$$M[1,6] = \frac{1}{(K_2 b)^2}\left((K_2 b)\frac{\zeta_l^{(2)\prime}(K_2 b)}{\zeta_l^{(2)}(K_2 b)}\frac{\zeta_l^{(2)}(K_2 b)}{\zeta_l^{(2)}(K_2 a)}\right)$$

$$M[1,7] = M[1,8] = 0$$

$$M[2,1] = -\frac{1}{(k_1 b)^2}\left(1 - (K_2 b)\frac{\zeta_l^{(1)\prime}(k_1 b)}{\zeta_l^{(1)}(k_1 b)}\right)$$

$$M[2,2] = -\frac{1}{(K_2 b)^2}(l(l+1))$$

$$M[2,3] = \frac{1}{(k_2 b)^2}\left(1 - (K_2 b)\frac{\zeta_l^{(1)\prime}(k_2 b)}{\zeta_l^{(1)}(k_2 b)}\right)\frac{\zeta_l^{(1)}(k_2 b)}{\zeta_l^{(1)}(k_2 a)}$$

$$M[2,4] = \frac{1}{(k_2b)^2}\left(1-(K_2b)\frac{\zeta_l^{(2)\prime}(k_2b)}{\zeta_l^{(2)}(k_2b)}\right)\frac{\zeta_l^{(2)}(k_2b)}{\zeta_l^{(2)}(k_2a)}$$

$$M[2,5] = \frac{1}{(K_2b)^2}\left(l(l+1)\right)\frac{\zeta_l^{(1)}(K_2b)}{\zeta_l^{(1)}(K_2a)}$$

$$M[2,6] = \frac{1}{(K_2b)^2}\left(l(l+1)\right)\frac{\zeta_l^{(2)}(K_2b)}{\zeta_l^{(2)}(K_2a)}$$

$$M[2,7] = M[2,8]=0$$

$$M[3,1] = -\frac{2\mu_1}{(k_1b)^2b}\left(2-(k_1b)\frac{\zeta_l^{(1)\prime}(k_1b)}{\zeta_l^{(1)}(k_1b)}\right)$$

$$M[3,2] = -\frac{2\mu_1}{(K_1b)^2b}\left(l(l+1)-\frac{(K_1b)^2}{2}-(K_1b)\frac{\zeta_l^{(1)\prime}(K_1b)}{\zeta_l^{(1)}(K_1b)}\right)$$

$$M[3,3] = \frac{2\mu_2}{(k_2b)^2b}\left(2-(k_2b)\frac{\zeta_l^{(1)\prime}(k_2b)}{\zeta_l^{(1)}(k_2b)}\right)\frac{\zeta_l^{(1)}(k_2b)}{\zeta_l^{(1)}(k_2a)}$$

$$M[3,4] = \frac{2\mu_2}{(k_2b)^2b}\left(2-(k_2b)\frac{\zeta_l^{(2)\prime}(k_2b)}{\zeta_l^{(2)}(k_2b)}\right)\frac{\zeta_l^{(2)}(k_2b)}{\zeta_l^{(2)}(k_2a)}$$

$$M[3,5] = \frac{2\mu_2}{(K_2b)^2b}\left(l(l+1)-\frac{(K_2b)^2}{2}-(K_2b)\frac{\zeta_l^{(1)\prime}(K_2b)}{\zeta_l^{(1)}(K_2b)}\right)\frac{\zeta_l^{(1)}(K_2b)}{\zeta_l^{(1)}(K_2a)}$$

$$M[3,6] = \frac{2\mu_2}{(K_2b)^2b}\left(l(l+1)-\frac{(K_2b)^2}{2}-(K_2b)\frac{\zeta_l^{(2)\prime}(K_2b)}{\zeta_l^{(2)}(K_2b)}\right)\frac{\zeta_l^{(2)}(K_2b)}{\zeta_l^{(2)}(K_2a)}$$

$$M[3,7] = M[3,8]=0$$

$$M[4,1] = -\frac{2\mu_1}{(k_1b)^2b}\left(\frac{(K_1b)^2}{2}-(l(l+1)+2)+2(k_1b)\frac{\zeta_l^{(1)\prime}(k_1b)}{\zeta_l^{(1)}(k_1b)}\right)$$

$$M[4,2] = \frac{2\mu_1}{(K_1b)^2b}\left(l(l+1)\left(2-(K_1b)\frac{\zeta_l^{(1)\prime}(K_1b)}{\zeta_l^{(1)}(K_1b)}\right)\right)$$

$$M[4,3] = \frac{2\mu_2}{(k_2b)^2 b}\left(\frac{(K_2b)^2}{2} - (l(l+1)+2) + 2(k_2b)\frac{\zeta_l^{(1)\prime}(k_2b)}{\zeta_l^{(1)}(k_2b)}\right)\frac{\zeta_l^{(1)}(k_2b)}{\zeta_l^{(1)}(k_2a)}$$

$$M[4,4] = \frac{2\mu_2}{(k_2b)^2 b}\left(\frac{(K_2b)^2}{2} - (l(l+1)+2) + 2(k_2b)\frac{\zeta_l^{(2)\prime}(k_2b)}{\zeta_l^{(2)}(k_2b)}\right)\frac{\zeta_l^{(2)}(k_2b)}{\zeta_l^{(2)}(k_2a)}$$

$$M[4,5] = -\frac{2\mu_2}{(K_2b)^2 b}\left(l(l+1)\left(2 - (K_2b)\frac{\zeta_l^{(1)\prime}(K_2b)}{\zeta_l^{(1)}(K_2b)}\right)\right)\frac{\zeta_l^{(1)}(K_2b)}{\zeta_l^{(1)}(K_2a)}$$

$$M[4,6] = -\frac{2\mu_2}{(K_2b)^2 b}\left(l(l+1)\left(2 - (K_2b)\frac{\zeta_l^{(2)\prime}(K_2b)}{\zeta_l^{(2)}(K_2b)}\right)\right)\frac{\zeta_l^{(2)}(K_2b)}{\zeta_l^{(2)}(K_2a)}$$

$$M[4,7] = M[4,8] = 0$$

$$M[5,1] = M[5,2] = 0$$

$$M[5,3] = \frac{2\mu_2}{(k_2a)^2 a}\left(\frac{(K_2a)^2}{2} - (l(l+1)+2) + 2(k_2a)\frac{\zeta_l^{(1)\prime}(k_2a)}{\zeta_l^{(1)}(k_2a)}\right)$$

$$M[5,4] = \frac{2\mu_2}{(k_2a)^2 a}\left(\frac{(K_2a)^2}{2} - (l(l+1)+2) + 2(k_2a)\frac{\zeta_l^{(2)\prime}(k_2a)}{\zeta_l^{(2)}(k_2a)}\right)$$

$$M[5,5] = -\frac{2\mu_2}{(K_2a)^2 a}\left(l(l+1)\left(2 - (K_2a)\frac{\zeta_l^{(1)\prime}(K_2a)}{\zeta_l^{(1)}(K_2a)}\right)\right)$$

$$M[5,6] = -\frac{2\mu_2}{(K_2a)^2 a}\left(l(l+1)\left(2 - (K_2a)\frac{\zeta_l^{(2)\prime}(K_2a)}{\zeta_l^{(2)}(K_2a)}\right)\right)$$

$$M[5,7] = \frac{2\mu_1}{(K_3a)^2 a}\left(l(l+1)\left(2 - (K_3a)\frac{\psi_l'(K_3a)}{\psi_l(K_3a)}\right)\right)$$

$$M[5,8] = -\frac{2\mu_3}{(k_3a)^2 a}\left(\frac{(K_3a)^2}{2} - (l(l+1)+2) + 2(k_3a)\frac{\psi_l'(k_3a)}{\psi_l(k_3a)}\right)$$

$$M[6,1] = M[6,2] = 0$$

$$M[6,3] = \frac{2\mu_2}{(k_2a)^2a}\left(2-(k_2a)\frac{\zeta_l^{(1)\prime}(k_2a)}{\zeta_l^{(1)}(k_2a)}\right)$$

$$M[6,4] = \frac{2\mu_2}{(k_2a)^2a}\left(2-(k_2a)\frac{\zeta_l^{(2)\prime}(k_2a)}{\zeta_l^{(2)}(k_2a)}\right)$$

$$M[6,5] = \frac{2\mu_2}{(K_2a)^2a}\left(l(l+1)-\frac{(K_2a)^2}{2}-(K_2a)\frac{\zeta_l^{(1)\prime}(K_2a)}{\zeta_l^{(1)}(K_2a)}\right)$$

$$M[6,6] = \frac{2\mu_2}{(K_2a)^2a}\left(l(l+1)-\frac{(K_2a)^2}{2}-(K_2a)\frac{\zeta_l^{(2)\prime}(K_2a)}{\zeta_l^{(2)}(K_2a)}\right)$$

$$M[6,7] = -\frac{2\mu_3}{(K_3a)^2a}\left(l(l+1)-\frac{(K_3a)^2}{2}-(K_3a)\frac{\psi_l'(K_3a)}{\psi_l(K_3a)}\right)$$

$$M[6,8] = -\frac{2\mu_3}{(k_3a)^2a}\left(2-(k_3a)\frac{\psi_l'(k_3a)}{\psi_l(k_3a)}\right)$$

$$M[7,1] = M[7,2]=0$$

$$M[7,3] = \frac{1}{(k_2a)^2}\left(1-(k_2a)\frac{\zeta_l^{(1)\prime}(k_2a)}{\zeta_l^{(1)}(k_2a)}\right)$$

$$M[7,4] = \frac{1}{(k_2a)^2}\left(1-(k_2a)\frac{\zeta_l^{(2)\prime}(k_2a)}{\zeta_l^{(2)}(k_2a)}\right)$$

$$M[7,5] = M[7,6]=\frac{1}{(K_2a)^2}\,(l(l+1))$$

$$M[7,7] = -\frac{1}{(K_3a)^2}\,(l(l+1))$$

$$M[7,8] = -\frac{1}{(k_3a)^2}\left(1-(k_3a)\frac{\psi_l'(k_3a)}{\psi_l(k_3a)}\right)$$

$$M[8,1] = M[8,2]=0$$

$$
\begin{aligned}
M[8,3] &= M[8,4] = -\frac{1}{(k_2a)^2} \\
M[8,5] &= \frac{1}{(K_2a)^2}\left((K_2a)\frac{\zeta_l^{(1)\prime}(K_2a)}{\zeta_l^{(1)}(K_2a)}\right) \\
M[8,6] &= \frac{1}{(K_2a)^2}\left((K_2a)\frac{\zeta_l^{(2)\prime}(K_2a)}{\zeta_l^{(2)}(K_2a)}\right) \\
M[8,7] &= -\frac{1}{(K_3a)^2}\left((K_3a)\frac{\psi_l'(K_3a)}{\psi_l(K_3a)}\right) \\
M[8,8] &= \frac{1}{(k_3a)^2}
\end{aligned}
\tag{5.6}
$$

We are now left with expressing the b column, representing the contributions of displacement and normal surface traction to the outer boundary due to the incident longitudinal displacement field. The b column vector has the following components after factoring out the radial function common to all terms:

$$
\begin{aligned}
b[1] &= \frac{1}{(k_1b)^2} \\
b[2] &= -\frac{1}{(k_1b)^2}\left(1-(K_2b)\frac{\psi_l'(k_1b)}{\psi_l(k_1b)}\right) \\
b[3] &= -\frac{2\mu_1}{(k_1b)^2b}\left(2-(k_1b)\frac{\psi_l'(k_1b)}{\psi_l(k_1b)}\right) \\
b[4] &= -\frac{2\mu_1}{(k_1b)^2b}\left(\frac{(K_1b)^2}{2}-(l(l+1)+2)+2(k_1b)\frac{\psi_l'(k_1b)}{\psi_l(k_1b)}\right) \\
b[5] &= b[6] = b[7] = b[8] = 0
\end{aligned}
\tag{5.7}
$$

The solution to the elastic scattering problem will be to write the coefficients of the displacement potentials in terms of the boundary conditions. The algebraic solution is given by the algebraic ratio of the determinants of the boundary value matrix and the substitution matrices. We define the determinants as $\Delta_i^{L,S,V}$, where $i = (0, 1, 2, .)$. The numbers in the expression for the determinant indicate the column of the matrix that b has been substituted into. Therefore, $x_1 = \frac{\Delta_1^L}{\Delta_0}$ will be the modal coefficient of the scattered longitudinal displacement potential by Cramer's rule. We are then left with solving for the coefficients in the potentials. The potentials of interest in the scattering of elastic waves are the scattered fields and we can write the modal coefficients of the scattered fields in terms of the determinants as:

$$
\begin{aligned}
a_{1l} &= -(2l+1)\frac{\psi_l(k_1 b)}{\zeta_l^{(1)}(k_1 b)}\left(\frac{\Delta_1^L}{\Delta_0}\right) \\
c_{1l} &= -(2l+1)\frac{\psi_l(k_1 b)}{\zeta_l^{(1)}(K_1 b)}\left(\frac{K_1}{k_1}\right)^2 \frac{P_l^{(0)}}{P_l^{(1)}\cos\phi}\left(\frac{\Delta_1^L}{\Delta_0}\right)
\end{aligned}
\tag{5.8}
$$

where the factors in front of the potentials are the ratio of terms factored out of the boundary condition matrix prior to taking the determinants.

5.2.1 Scattered fields

The scattered fields will be both longitudinal and transverse. The potentials are given, upon substitution of the expressions for the coefficients, as:

$$
(k_1 r)\pi_L^s = \sum_{l=0}^{\infty} -(2l+1)\frac{\psi_l(k_1 b)}{\zeta_l^{(1)}(k_1 b)}\left(\frac{\Delta_1^L}{\Delta_0}\right) i^{l+1}\zeta_l^{(1)}(k_1 r)P_l(\cos\theta)
$$

$$(K_1 r)\pi_V^s = \sum_{l=1}^{\infty} -(2l+1)\frac{\psi_l(k_1 b)}{\zeta_l^{(1)}(K_1 b)}\left(\frac{K_1}{k_1}\right)^2 P_l^{(0)}\left(\frac{\Delta_1^L}{\Delta_0}\right) i^{l-1}\zeta_l^{(1)}(K_1 r) \tag{5.9}$$

We can see from the displacement potentials that the coupling of these scattered displacement fields to the displacement fields inside of the interface layer and in the interior are completely contained within the determinant ratios. All other factors in the displacement potentials are dependent only on the external field variables. If we take a look at the expression for Δ_0 (equation 5.10) there is a clear structure to the algebraic solution that can be seen. First we should note that the subscripts on the components of the factors in the determinant actually represent the matrix element position after extensive simplification of the boundary condition matrix. Secondly, we note that through careful manipulation, we have not mixed the expressions for the interior field with those of the exterior field in the matrix manipulations. The first four terms in Δ_0 are coupling terms between the exterior field and interface layer, and the remaining four terms are coupling between the interface layer and the interior. This interface coupling, is a result of the exterior and interior fields being separated by the interface layer, the dependence of interior and exterior fields must be through the properties of the interface layer. The expressions for $\Delta_0, \Delta_1^L, \Delta_2^L$ are given below, with the last four elements of the determinant identical for all three determinants. This is expected, since the relations between the interface layer and interior region must not be affected by substitution of the column vector b into the columns corresponding to components from the scattered field. The determinants are:

$$\Delta_0 = Mult_0(1 - C_{11}C_{88} - C_{12}C_{78} - C_{21}C_{87} - C_{22}C_{77} + (C_{11}C_{22} - C_{21}C_{21})(C_{77}C_{88} - C_{87}C_{78})) \tag{5.10}$$

$$C_{11} = \frac{1}{\frac{\zeta_l^{(2)}(K_2b)}{\zeta_l^{(2)}(K_2a)}\left(F_2^T - \frac{(K_2b)\zeta_l^{(2)\prime}(K_2b)}{\zeta_l^{(2)}(K_2b)}\right)}\left[1 + \frac{\frac{1}{2}(K_2b)^2\frac{\mu_2}{\mu_2-\mu_1}\left(-l(l+1) - \frac{1}{2}(K_1b)^2\frac{\mu_1}{\mu_2-\mu_1} + \frac{(k_1b)h_l^{(1)\prime}(k_2b)}{h_l^{(1)}(k_2b)}\frac{(K_1b)\zeta_l^{(1)\prime}(K_1b)}{\zeta_l^{(1)}(K_1b)}\right)}{Den_1}\right]$$

$$C_{12} = \frac{1}{\frac{\zeta_l^{(2)}(K_2b)}{\zeta_l^{(2)}(K_2a)}\left(F_2^T - \frac{(K_2b)\zeta_l^{(2)\prime}(K_2b)}{\zeta_l^{(2)}(K_2b)}\right)}\left[F_2^T + \frac{\frac{1}{2}(K_2b)^2\frac{\mu_2}{\mu_2-\mu_1}\left(-2l(l+1) + 2\frac{(k_1b)h_l^{(1)\prime}(k_2b)}{h_l^{(1)}(k_2b)}\frac{(K_1b)\zeta_l^{(1)\prime}(K_1b)}{\zeta_l^{(1)}(K_1b)} - \frac{1}{2}(K_1b)^2\frac{\mu_1}{\mu_2-\mu_1}\frac{(K_1b)\zeta_l^{(1)\prime}(K_1b)}{\zeta_l^{(1)}(K_1b)}\right)}{Den_1}\right]$$

$$C_{21} = \frac{1}{\frac{\zeta_l^{(2)}(k_2b)}{\zeta_l^{(2)}(k_2a)}\left(F_2^L - \frac{(k_2b)\zeta_l^{(2)\prime}(k_2b)}{\zeta_l^{(2)}(k_2b)}\right)}\left[F_2^L + \frac{\frac{1}{2}(K_2b)^2\frac{\mu_2}{\mu_2-\mu_1}\left(-l(l+1) - \frac{(k_1b)h_l^{(1)\prime}(k_2b)}{h_l^{(1)}(k_2b)}\frac{1}{2}(K_1b)^2\frac{\mu_1}{\mu_2-\mu_1} + \frac{(k_1b)h_l^{(1)\prime}(k_2b)}{h_l^{(1)}(k_2b)}\frac{(K_1b)\zeta_l^{(1)\prime}(K_1b)}{\zeta_l^{(1)}(K_1b)}\right)}{Den_1}\right]$$

$$C_{22} = \frac{LL1}{\frac{\zeta_l^{(2)}(k_2b)}{\zeta_l^{(2)}(k_2a)}\left(F_2^L - \frac{(k_2b)\zeta_l^{(2)\prime}(k_2b)}{\zeta_l^{(2)}(k_2b)}\right)}\left[1 + \frac{\frac{1}{2}(K_2b)^2\frac{\mu_2}{\mu_2-\mu_1}\left(-l(l+1) - \frac{1}{2}(K_1b)^2\frac{\mu_1}{\mu_2-\mu_1} + \frac{(k_1b)h_l^{(1)\prime}(k_2b)}{h_l^{(1)}(k_2b)}\frac{(K_1b)\zeta_l^{(1)\prime}(K_1b)}{\zeta_l^{(1)}(K_1b)}\right)}{CD_1}\right]$$

$$C_{77} = \frac{-LL1}{F_1^L}\left[1 + \frac{\frac{1}{2}(K_2a)^2\frac{\mu_2}{\mu_2-\mu_3}\left(-l(l+1) - \frac{(k_3a)j_l'(k_3a)}{j_l(k_3a)}\frac{(K_3a)\psi_l'(K_3a)}{\psi_l(K_3a)} - \frac{1}{2}(K_3a)^2\frac{\mu_3}{\mu_2-\mu_3}\right)}{CD_2}\right]$$

$$C_{78} = \frac{-1}{F_1^L}\left[\frac{\frac{(k_2a)h_l^{(2)\prime}(k_2a)}{h_l^{(2)}(k_2a)} - F_1^L\frac{\zeta_l^{(2)}(k_2b)}{\zeta_l^{(2)}(k_2a)} + \frac{1}{2}(K_2a)^2\frac{\mu_2}{\mu_2-\mu_3}\left(-l(l+1) + \frac{(k_3a)j_l'(k_3a)}{j_l(k_3a)}\frac{(K_3a)\psi_l'(K_3a)}{\psi_l(K_3a)} - \frac{(k_3a)j_l'(k_3a)}{j_l(k_3a)}\frac{1}{2}(K_3a)^2\frac{\mu_3}{\mu_2-\mu_3}\right)}{Den_2}\right]$$

$$C_{87} = \frac{-1}{F_1^T}\left[\frac{\frac{(K_2a)\zeta_l^{(2)\prime}(K_2a)}{\zeta_l^{(2)}(K_2a)} - F_1^T\frac{\zeta_l^{(2)}(k_2b)}{\zeta_l^{(2)}(k_2a)} + \frac{1}{2}(K_2a)^2\frac{\mu_2}{\mu_2-\mu_3}\left(-2l(l+1) + 2\frac{(k_3a)j_l'(k_3a)}{j_l(k_3a)}\frac{(K_3a)\psi_l'(K_3a)}{\psi_l(K_3a)} - \frac{(K_3a)\psi_l'(K_3a)}{\psi_l(K_3a)}\frac{1}{2}(K_3a)^2\frac{\mu_3}{\mu_2-\mu_3}\right)}{Den_2}\right]$$

$$C_{88} = \frac{-1}{F_1^T}\left[1 + \frac{\frac{1}{2}(K_2a)^2\frac{\mu_2}{\mu_2-\mu_3}\left(-l(l+1) + \frac{(k_3a)j_l'(k_3a)}{j_l(k_3a)}\frac{(K_3a)\psi_l'(K_3a)}{\psi_l(K_3a)} - \frac{1}{2}(K_3a)^2\frac{\mu_3}{\mu_2-\mu_3}\right)}{Den_2}\right]$$

$$\begin{aligned} Den_1 &= LL1\left(1-\frac{(k_1b)h_l^{(1)\prime}(k_2b)}{h_l^{(1)}(k_2b)}\right)\left(-2+\frac{(K_1b)\zeta_l^{(1)\prime}(K_1b)}{\zeta_l^{(1)}(K_1b)}\right) \\ &+\left(2\frac{(k_1b)h_l^{(1)\prime}(k_2b)}{h_l^{(1)}(k_2b)}-LL1-\frac{1}{2}(K_1b)^2\frac{\mu_1}{\mu_2-\mu_1}\right)\left(-LL1-\frac{1}{2}(K_1b)^2\frac{\mu_1}{\mu_2-\mu_1}+\frac{(K_1b)\zeta_l^{(1)\prime}(K_1b)}{\zeta_l^{(1)}(K_1b)}\right) \end{aligned}$$

$$\begin{aligned} Den_2 &= LL1\left(1-\frac{(k_3a)j_l'(k_3a)}{j_l(k_3a)}\right)\left(-2+\frac{(K_3a)\psi_l'(K_3a)}{\psi_l(K_3a)}\right) \\ &+\left(2\frac{(k_3a)j_l'(k_3a)}{j_l(k_3a)}-LL1-\frac{1}{2}(K_3a)^2\frac{\mu_3}{\mu_2-\mu_3}\right)\left(-LL1+\frac{(K_3a)\psi_l'(K_3a)}{\psi_l(K_3a)}-\frac{1}{2}(K_3a)^2\frac{\mu_3}{\mu_2-\mu_3}\right) \end{aligned}$$

$$\begin{aligned} Mult_0 &= \left(\frac{\zeta_l^{(1)}(k_2b)}{\zeta_l^{(1)}(k_2a)}-\frac{\zeta_l^{(2)}(k_2b)}{\zeta_l^{(2)}(k_2a)}\right)\left(\frac{\zeta_l^{(1)}(K_2b)}{\zeta_l^{(1)}(K_2a)}-\frac{\zeta_l^{(2)}(K_2b)}{\zeta_l^{(2)}(K_2a)}\right)\left(\frac{\mu2-\mu1}{\mu1}\right)^2\left(\frac{\mu2-\mu3}{\mu3}\right)^2 \\ &\left(F_1^L F_1^T\frac{\zeta_l^{(2)}(k_2b)}{\zeta_l^{(2)}(k_2a)}\frac{\zeta_l^{(2)}(K_2b)}{\zeta_l^{(2)}(K_2a)}(F_2^L-\frac{(k_2b)\zeta_l^{(2)\prime}(k_2b)}{\zeta_l^{(2)}(k_2b)})(F_2^T-\frac{(K_2b)\zeta_l^{(2)\prime}(K_2b)}{\zeta_l^{(2)}(K_2b)})Den_1Den_2\right) \end{aligned}$$

$$F_1^L=\frac{(k_2a)\left(\zeta_l^{(1)}(k_2a)\zeta_l^{(2)}(k_2a)\right)'}{\left(\zeta_l^{(1)}(k_2b)\zeta_l^{(2)}(k_2a)-\zeta_l^{(2)}(k_2b)\zeta_l^{(1)}(k_2a)\right)}\qquad F_2^L=\frac{(k_2b)\left(\zeta_l^{(1)\prime}(k_2b)\zeta_l^{(2)}(k_2a)-\zeta_l^{(2)\prime}(k_2b)\zeta_l^{(1)}(k_2a)\right)}{\left(\zeta_l^{(1)}(k_2b)\zeta_l^{(2)}(k_2a)-\zeta_l^{(2)}(k_2b)\zeta_l^{(1)}(k_2a)\right)}$$

$$F_1^T=\frac{(K_2a)\left(\zeta_l^{(1)}(K_2a)\zeta_l^{(2)}(K_2a)\right)'}{\left(\zeta_l^{(1)}(K_2b)\zeta_l^{(2)}(K_2a)-\zeta_l^{(2)}(K_2b)\zeta_l^{(1)}(K_2a)\right)}\qquad F_2^T=\frac{(K_2b)\left(\zeta_l^{(1)\prime}(K_2b)\zeta_l^{(2)}(K_2a)-\zeta_l^{(2)\prime}(K_2b)\zeta_l^{(1)}(K_2a)\right)}{\left(\zeta_l^{(1)}(K_2b)\zeta_l^{(2)}(K_2a)-\zeta_l^{(2)}(K_2b)\zeta_l^{(1)}(K_2a)\right)}$$

$$\Delta_1^L = Mult_1^L\,(1 - C_{11}C_{88} - C_{12}C_{78} - C_{21}C_{87} - C_{22}C_{77} + (C_{11}C_{22} - C_{21}C_{21})(C_{77}C_{88} - C_{87}C_{78})) \tag{5.11}$$

$$C_{11} = -\frac{1}{\frac{\zeta_l^{(2)}(K_2b)}{\zeta_l^{(2)}(K_2a)}(F_2^T - \frac{(K_2b)\zeta_l^{(2)\prime}(K_2b)}{\zeta_l^{(2)}(K_2b)})}\left(F_2^T - \frac{\frac{1}{2}(K_2b)^2\frac{\mu_2}{\mu_2-\mu_1}(2l(l+1) - \frac{(K_1b)\zeta_l^{(1)\prime}(K_1b)}{\zeta_l^{(1)}(K_1b)}(2\frac{(k_1b)j_l^{(1)\prime}(k_2b)}{j_l^{(1)}(k_2b)} - \frac{1}{2}(K_1b)^2\frac{\mu_1}{\mu_2-\mu_1}))}{CD_1}\right)$$

$$C_{12} = \frac{1}{(\frac{\zeta_l^{(2)}(K_2b)}{\zeta_l^{(2)}(K_2a)}(F_2^T - \frac{(K_2b)\zeta_l^{(2)\prime}(K_2b)}{\zeta_l^{(2)}(K_2b)}))}\left(1 + \frac{\frac{1}{2}(K_2b)^2\frac{\mu_2}{\mu_2-\mu_1}(l(l+1) - \frac{(k_1b)j_l^{(1)\prime}(k_2b)}{j_l^{(1)}(k_2b)}\frac{(K_1b)\zeta_l^{(1)\prime}(K_1b)}{\zeta_l^{(1)}(K_1b)} + \frac{1}{2}(K_1b)^2\frac{\mu_1}{\mu_2-\mu_1})}{CD_1}\right)$$

$$C_{21} = \frac{-l(l+1)}{(\frac{\zeta_l^{(2)}(k_2b)}{\zeta_l^{(2)}(k_2a)}(F_2^L - \frac{(k_2b)\zeta_l^{(2)\prime}(k_2b)}{\zeta_l^{(2)}(k_2b)}))}\left(1 + \frac{\frac{1}{2}(K_2b)^2\frac{\mu_2}{\mu_2-\mu_1}(l(l+1) - \frac{(k_1b)j_l^{(1)\prime}(k_2b)}{j_l^{(1)}(k_2b)}\frac{(K_1b)\zeta_l^{(1)\prime}(K_1b)}{\zeta_l^{(1)}(K_1b)} + \frac{1}{2}(K_1b)^2\frac{\mu_1}{\mu_2-\mu_1})}{CD_1}\right)$$

$$C_{22} = \frac{1}{\frac{\zeta_l^{(2)}(k_2b)}{\zeta_l^{(2)}(k_2a)}(F_2^L - \frac{(k_2b)\zeta_l^{(2)\prime}(k_2b)}{\zeta_l^{(2)}(k_2b)})}\left(F_2^L - 1 + \frac{\frac{1}{2}(K_2b)^2\frac{\mu_2}{\mu_2-\mu_1}(l(l+1) - \frac{(k_1b)j_l^{(1)\prime}(k_2b)}{j_l^{(1)}(k_2b)}(\frac{(K_1b)\zeta_l^{(1)\prime}(K_1b)}{\zeta_l^{(1)}(K_1b)} - \frac{1}{2}(K_1b)^2\frac{\mu_1}{\mu_2-\mu_1}))}{CD_1}\right)$$

$$\begin{aligned} Mult_1^L &= \left(\frac{\zeta_l^{(1)}(k_2b)}{\zeta_l^{(1)}(k_2a)} - \frac{\zeta_l^{(2)}(k_2b)}{\zeta_l^{(2)}(k_2a)}\right)\left(\frac{\zeta_l^{(1)}(K_2b)}{\zeta_l^{(1)}(K_2a)} - \frac{\zeta_l^{(2)}(K_2b)}{\zeta_l^{(2)}(K_2a)}\right)\left(\frac{\mu 2 - \mu 1}{\mu 1}\right)^2\left(\frac{\mu 2 - \mu 3}{\mu 3}\right)^2 \\ &\quad \left(F_1^L F_1^T \frac{\zeta_l^{(2)}(k_2b)}{\zeta_l^{(2)}(k_2a)}\frac{\zeta_l^{(2)}(K_2b)}{\zeta_l^{(2)}(K_2a)}(F_2^L - \frac{(k_2b)\zeta_l^{(2)\prime}(k_2b)}{\zeta_l^{(2)}(k_2b)})(F_2^T - \frac{(K_2b)\zeta_l^{(2)\prime}(K_2b)}{\zeta_l^{(2)}(K_2b)})CD_1CD_2\right) \end{aligned}$$

$$\begin{aligned} CD_1 &= l(l+1)\left(2\left(1 - \frac{1}{2}(K_1b)^2\frac{\mu_1}{\mu_2-\mu_1}\right) - l(l+1) + \frac{(k_1b)j_l^{(1)\prime}(k_2b)}{j_l^{(1)}(k_2b)}\frac{(K_1b)\zeta_l^{(1)\prime}(K_1b)}{\zeta_l^{(1)}(K_1b)}\right) \\ &\quad -2\frac{(k_1b)j_l^{(1)\prime}(k_2b)}{j_l^{(1)}(k_2b)}\frac{(K_1b)\zeta_l^{(1)\prime}(K_1b)}{\zeta_l^{(1)}(K_1b)} + \frac{1}{2}(K_1b)^2\frac{\mu_1}{\mu_2-\mu_1}\left(2\frac{(k_1b)j_l^{(1)\prime}(k_2b)}{j_l^{(1)}(k_2b)} - \frac{1}{2}(K_1b)^2\frac{\mu_1}{\mu_2-\mu_1} + \frac{(K_1b)\zeta_l^{(1)\prime}(K_1b)}{\zeta_l^{(1)}(K_1b)}\right) \end{aligned}$$

$$\Delta_2^L = \boldsymbol{Mult_2^L}\,(1 - C_{11}C_{88} - C_{12}C_{78} - C_{21}C_{87} - C_{22}C_{77} + (C_{11}C_{22} - C_{21}C_{21})(C_{77}C_{88} - C_{87}C_{78})) \tag{5.12}$$

$$C_{11} = -\frac{1}{\frac{\zeta_l^{(2)}(K_2b)}{\zeta_l^{(2)}(K_2a)}(F_2^T - \frac{(K_2b)\zeta_l^{(2)\prime}(K_2b)}{\zeta_l^{(2)}(K_2b)})}\left(F_2^T - \frac{2\frac{1}{2}(K_2b)^2\frac{\mu_2}{\mu_2-\mu_1}}{CD_1}\right)$$

$$C_{12} = \frac{1}{(\frac{\zeta_l^{(2)}(K_2b)}{\zeta_l^{(2)}(K_2a)}(F_2^T - \frac{(K_2b)\zeta_l^{(2)\prime}(K_2b)}{\zeta_l^{(2)}(K_2b)}))}\left(-1 + \frac{\frac{1}{2}(K_2b)^2\frac{\mu_2}{\mu_2-\mu_1}}{CD_1}\right)$$

$$C_{21} = \frac{-l(l+1)}{(\frac{\zeta_l^{(2)}(k_2b)}{\zeta_l^{(2)}(k_2a)}(F_2^L - \frac{(k_2b)\zeta_l^{(2)\prime}(k_2b)}{\zeta_l^{(2)}(k_2b)}))}\left(-1 + \frac{\frac{1}{2}(K_2b)^2\frac{\mu_2}{\mu_2-\mu_1}(l(l+1) + \frac{1}{2}(K_1b)^2\frac{\mu_1}{\mu_2-\mu_1})}{CD_1}\right)$$

$$C_{22} = \frac{1}{\frac{\zeta_l^{(2)}(k_2b)}{\zeta_l^{(2)}(k_2a)}(F_2^L - \frac{(k_2b)\zeta_l^{(2)\prime}(k_2b)}{\zeta_l^{(2)}(k_2b)})}\left(F_2^L - 1 + \frac{\frac{1}{2}(K_2b)^2\frac{\mu_2}{\mu_2-\mu_1}}{CD_1}\right)$$

$$CD_1 = -\left(2 - \frac{1}{2}(K_1b)^2\frac{\mu_1}{\mu_2-\mu_1} - l(l+1)\right)$$

$$\begin{aligned}\boldsymbol{Mult_2^L} = {} & \left(\frac{\zeta_l^{(1)}(k_2b)}{\zeta_l^{(1)}(k_2a)} - \frac{\zeta_l^{(2)}(k_2b)}{\zeta_l^{(2)}(k_2a)}\right)\left(\frac{\zeta_l^{(1)}(K_2b)}{\zeta_l^{(1)}(K_2a)} - \frac{\zeta_l^{(2)}(K_2b)}{\zeta_l^{(2)}(K_2a)}\right)\left(\frac{\mu_2-\mu_1}{\mu_1}\right)^2\left(\frac{\mu_2-\mu_3}{\mu_3}\right)^2 CD_1CD_2 \\ & \left(F_1^LF_1^T\frac{\zeta_l^{(2)}(k_2b)}{\zeta_l^{(2)}(k_2a)}\frac{\zeta_l^{(2)}(K_2b)}{\zeta_l^{(2)}(K_2a)}(F_2^L - \frac{(k_2b)\zeta_l^{(2)\prime}(k_2b)}{\zeta_l^{(2)}(k_2b)})(F_2^T - \frac{(K_2b)\zeta_l^{(2)\prime}(K_2b)}{\zeta_l^{(2)}(K_2b)})\right)\left(\frac{(k_1b)h_l^{(1)\prime}(k_2b)}{h_l^{(1)}(k_2b)} - \frac{(k_1b)j_l^{(1)\prime}(k_2b)}{j_l^{(1)}(k_2b)}\right)\end{aligned}$$

5.2.2 Dynamic stress concentrations

The expressions for the scattered displacement potentials allow us to explicitly write the expressions for the dynamic stress concentrations. These can then be evaluated to determine the effect of an interface layer. We normalize the dynamic stress concentrations by calculating the incident stress field and computing a dimensionless ratio σ^*, that will be zero for no scattered stress field. There will be three terms adding to σ^* and we will review the individual contributions in order to interpret the physics of the problem. The radial stresses are given as:

$$
\begin{aligned}
\sigma^*_{rr} &= \frac{\left|\sigma^L_{rr} + \sigma^V_{rr}\right|}{\left|\sigma^i_{xx}\right|} \\
\sigma^L_{rr} &= \sum_{n=0}^{\infty} \frac{2\mu_1}{(k_1 r)^2 r} \left(\frac{(K_1 r)^2}{2} - (l(l+1)+2) + 2(k_1 r)\frac{\zeta_l^{(1)\prime}(k_1 r)}{\zeta_l^{(1)}(k_1 r)} \right) \\
& \quad (2l+1)\frac{\psi_l(k_1 b)}{\zeta_l^{(1)}(k_1 b)} \left(\frac{\Delta_1^L}{\Delta_0}\right) i^{l+1} \zeta_l^{(1)}(k_1 r) P_l(\cos\theta) \\
\sigma^V_{rr} &= \sum_{n=0}^{\infty} \frac{2\mu_1}{(K_1 r)^2 r} \left(l(l+1)\left(2 - (K_1 r)\frac{\zeta_l^{(1)\prime}(K_1 r)}{\zeta_l^{(1)}(K_1 r)}\right)\right) \\
& \quad (2l+1)\frac{\psi_l(k_1 b)}{\zeta_l^{(1)}(K_1 b)} \left(\frac{K_1}{k_1}\right)^2 P_l^{(0)} \left(\frac{\Delta_1^L}{\Delta_0}\right) i^{l-1} \zeta_l^{(1)}(K_1 r) \qquad (5.13)
\end{aligned}
$$

with the shear stresses:

$$
\begin{aligned}
\sigma^*_{r\theta} &= \frac{\left|\sigma^L_{r\theta} + \sigma^V_{r\theta}\right|}{\left|\sigma^i_{xx}\right|} \\
\sigma^L_{r\theta} &= \sum_{n=0}^{\infty} \frac{2\mu_1}{(k_1 r)^2 r} \left(2 - (k_1 r)\frac{\zeta_l^{(1)\prime}(k_1 r)}{\zeta_l^{(1)}(k_1 r)} \right)
\end{aligned}
$$

$$(2l+1)\frac{\psi_l(k_1b)}{\zeta_l^{(1)}(k_1b)}\left(\frac{\Delta_1^L}{\Delta_0}\right)i^{l+1}\zeta_l^{(1)}(k_1r)P_l(\cos\theta)$$

$$\sigma_{r\theta}^V = \sum_{n=0}^{\infty}-\frac{2\mu_1}{(K_1r)^2r}\left(l(l+1)-\frac{(K_1r)^2}{2}-(K_1r)\frac{\zeta_l^{(1)\prime}(K_1r)}{\zeta_l^{(1)}(K_1r)}\right)$$

$$(2l+1)\frac{\psi_l(k_1b)}{\zeta_l^{(1)}(K_1b)}\left(\frac{K_1}{k_1}\right)^2 P_l^{(0)}\left(\frac{\Delta_1^L}{\Delta_0}\right)i^{l-1}\zeta_l^{(1)}(K_1r) \tag{5.14}$$

5.3 Incident coupled transverse wave

For an incident transverse wave of the π_V type, we have for the incident potential field

$$(K_1r)\pi_V^i = \sum_{l=1}^{\infty} i^{l-1}\left(\frac{2l+1}{l(l+1)}\right)\psi_l(K_1r)P_l^{(1)}(\cos\theta)\cos\phi \tag{5.15}$$

The same boundary conditions are needed to solve for the scattered fields as in the incident longitudinal case. We will need to solve the 8×8 system to calculate the coefficients of the potentials. The matrix M is unchanged, but the column vector b, which represents the incident field components of displacement and normal surface traction on the boundaries, is modified. The b column vector for the incident transverse wave has the components:

$$b[1] = -\frac{1}{(K_1b)^2}\left((K_1b)\frac{\psi_l'(K_1b)}{\psi_l(K_1b)}\right)$$

$$b[2] = -\frac{1}{(K_2b)^2}(l(l+1))$$

$$b[3] = -\frac{2\mu_1}{(K_1b)^2b}\left(l(l+1)-\frac{(K_1b)^2}{2}-(K_1b)\frac{\psi_l'(K_1b)}{\psi_l(K_1b)}\right)$$

$$b[4] = \frac{2\mu_1}{(K_1b)^2b}\left(l(l+1)\left(2-(K_1b)\frac{\psi_l'(K_1b)}{\psi_l(K_1b)}\right)\right)$$

$$b[5] = b[6]=b[7]=b[8]=0 \tag{5.16}$$

We have factored out the radial functions in the incident field. We now substitute the column vector b into the matrix M and recompute the determinants, solving for the unknown modal coefficients A_{1l} and B_{1l}. The expression for Δ_0 remains unchanged from the previously given expression and the coefficients of the scattered fields can be written as:

$$\begin{aligned} A_{1l} &= -\frac{(2l+1)}{l(l+1)} \frac{\psi_l(K_1 b)}{\zeta_l^{(1)}(k_1 b)} \left(\frac{k_1}{K_1}\right)^2 \frac{P_l^{(1)} \cos\phi}{P_l^{(0)}} \left(\frac{\Delta_1^V}{\Delta_0}\right) \\ B_{1l} &= \frac{(2l+1)}{l(l+1)} \frac{\psi_l(K_1 b)}{\zeta_l^{(1)}(K_1 b)} \left(\frac{\Delta_1^V}{\Delta_0}\right) \end{aligned} \quad (5.17)$$

where the factors in front of the potentials are the ratio of terms factored out of the boundary condition matrix prior to taking the determinants.

5.3.1 Scattered fields

The scattered fields due to the incident transverse displacement field will be both longitudinal and transverse. The potentials are given, upon substitution of the expressions for the coefficients, as:

$$\begin{aligned} (k_1 r)\pi_L^s &= \sum_{l=0}^{\infty} -\frac{(2l+1)}{l(l+1)} \frac{\psi_l(K_1 b)}{\zeta_l^{(1)}(k_1 b)} \left(\frac{k_1}{K_1}\right)^2 P_l^{(1)} \cos\phi \left(\frac{\Delta_1^V}{\Delta_0}\right) i^{l+1} \zeta_l^{(1)}(k_1 r) \\ (K_1 r)\pi_V^s &= \sum_{l=1}^{\infty} \frac{(2l+1)}{l(l+1)} \frac{\psi_l(K_1 b)}{\zeta_l^{(1)}(K_1 b)} \left(\frac{\Delta_1^V}{\Delta_0}\right) P_l^{(1)} \cos\phi\, i^{l-1} \zeta_l^{(1)}(K_1 r) \end{aligned} \quad (5.18)$$

with the expressions for Δ_1^V and Δ_2^V given by:

$$\Delta_1^V = Mult_1^V \left(1 - C_{11}C_{88} - C_{12}C_{78} - C_{21}C_{87} - C_{22}C_{77} + (C_{11}C_{22} - C_{21}C_{21})(C_{77}C_{88} - C_{87}C_{78})\right) \tag{5.19}$$

$$\begin{aligned}
C_{11} &= -\frac{1}{\frac{\zeta_l^{(2)}(K_2b)}{\zeta_l^{(2)}(K_2a)}\left(F_2^T - \frac{(K_2b)\zeta_l^{(2)\prime}(K_2b)}{\zeta_l^{(2)}(K_2b)}\right)} \left(F_2^T - \frac{2\frac{1}{2}\,(K_2b)^2\,\frac{\mu_2}{\mu_2-\mu_1}}{CD_1}\right) \\
C_{12} &= \frac{1}{\left(\frac{\zeta_l^{(2)}(K_2b)}{\zeta_l^{(2)}(K_2a)}\left(F_2^T - \frac{(K_2b)\zeta_l^{(2)\prime}(K_2b)}{\zeta_l^{(2)}(K_2b)}\right)\right)} \left(-1 + \frac{\frac{1}{2}\,(K_2b)^2\,\frac{\mu_2}{\mu_2-\mu_1}}{CD_1}\right) \\
C_{21} &= \frac{-l\,(l+1)}{\left(\frac{\zeta_l^{(2)}(k_2b)}{\zeta_l^{(2)}(k_2a)}\left(F_2^L - \frac{(k_2b)\zeta_l^{(2)\prime}(k_2b)}{\zeta_l^{(2)}(k_2b)}\right)\right)} \left(-1 + \frac{\frac{1}{2}\,(K_2b)^2\,\frac{\mu_2}{\mu_2-\mu_1}\left(l\,(l+1) + \frac{1}{2}\,(K_1b)^2\,\frac{\mu_1}{\mu_2-\mu_1}\right)}{CD_1}\right) \\
C_{22} &= \frac{1}{\frac{\zeta_l^{(2)}(k_2b)}{\zeta_l^{(2)}(k_2a)}\left(F_2^L - \frac{(k_2b)\zeta_l^{(2)\prime}(k_2b)}{\zeta_l^{(2)}(k_2b)}\right)} \left(F_2^L - 1 + \frac{\frac{1}{2}\,(K_2b)^2\,\frac{\mu_2}{\mu_2-\mu_1}}{CD_1}\right)
\end{aligned}$$

$$CD_1 = -\left(2 - \frac{1}{2}(K_1b)^2\,\frac{\mu_1}{\mu_2-\mu_1} - l\,(l+1)\right)$$

$$\begin{aligned}
Mult_1^V = {} & \left(\frac{\zeta_l^{(1)}(k_2b)}{\zeta_l^{(1)}(k_2a)} - \frac{\zeta_l^{(2)}(k_2b)}{\zeta_l^{(2)}(k_2a)}\right)\left(\frac{\zeta_l^{(1)}(K_2b)}{\zeta_l^{(1)}(K_2a)} - \frac{\zeta_l^{(2)}(K_2b)}{\zeta_l^{(2)}(K_2a)}\right)\left(\frac{\mu_2-\mu_1}{\mu_1}\right)^2\left(\frac{\mu_2-\mu_3}{\mu_3}\right)^2 CD_1CD_2 \\
& \left(F_1^LF_1^T\frac{\zeta_l^{(2)}(k_2b)}{\zeta_l^{(2)}(k_2a)}\frac{\zeta_l^{(2)}(K_2b)}{\zeta_l^{(2)}(K_2a)}\left(F_2^L - \frac{(k_2b)\zeta_l^{(2)\prime}(k_2b)}{\zeta_l^{(2)}(k_2b)}\right)\left(F_2^T - \frac{(K_2b)\zeta_l^{(2)\prime}(K_2b)}{\zeta_l^{(2)}(K_2b)}\right)\right)\left(\frac{(K_1b)j_l^{(1)\prime}(K_2b)}{j_l^{(1)}(K_2b)} - \frac{(K_1b)h_l^{(1)\prime}(K_2b)}{h_l^{(1)}(K_2b)}\right)
\end{aligned}$$

$$\Delta_2^V = Mult_2^V\,(1 - C_{11}C_{88} - C_{12}C_{78} - C_{21}C_{87} - C_{22}C_{77} + (C_{11}C_{22} - C_{21}C_{21})(C_{77}C_{88} - C_{87}C_{78})) \tag{5.20}$$

$$C_{11} = -\frac{1}{\frac{\zeta_l^{(2)}(K_2b)}{\zeta_l^{(2)}(K_2a)}(F_2^T - \frac{(K_2b)\zeta_l^{(2)\prime}(K_2b)}{\zeta_l^{(2)}(K_2b)})}\left(F_2^T - \frac{\frac{1}{2}(K_2b)^2\frac{\mu_2}{\mu_2-\mu_1}(2l(l+1) - \frac{(K_1b)\psi_l^{(1)\prime}(K_1b)}{\psi_l^{(1)}(K_1b)}\,(2\frac{(k_1b)h_l^{(1)\prime}(k_2b)}{h_l^{(1)}(k_2b)} - \frac{1}{2}(K_1b)^2\frac{\mu_1}{\mu_2-\mu_1}))}{CD_1}\right)$$

$$C_{12} = \frac{1}{(\frac{\zeta_l^{(2)}(K_2b)}{\zeta_l^{(2)}(K_2a)}(F_2^T - \frac{(K_2b)\zeta_l^{(2)\prime}(K_2b)}{\zeta_l^{(2)}(K_2b)}))}\left(1 + \frac{\frac{1}{2}(K_2b)^2\frac{\mu_2}{\mu_2-\mu_1}(l(l+1) - \frac{(k_1b)h_l^{(1)\prime}(k_2b)}{h_l^{(1)}(k_2b)}\frac{(K_1b)\psi_l^{(1)\prime}(K_1b)}{\psi_l^{(1)}(K_1b)} + \frac{1}{2}(K_1b)^2\frac{\mu_1}{\mu_2-\mu_1})}{CD_1}\right)$$

$$C_{21} = \frac{-l(l+1)}{(\frac{\zeta_l^{(2)}(k_2b)}{\zeta_l^{(2)}(k_2a)}(F_2^L - \frac{(k_2b)\zeta_l^{(2)\prime}(k_2b)}{\zeta_l^{(2)}(k_2b)}))}\left(1 + \frac{\frac{1}{2}(K_2b)^2\frac{\mu_2}{\mu_2-\mu_1}(l(l+1) - \frac{(k_1b)h_l^{(1)\prime}(k_2b)}{h_l^{(1)}(k_2b)}\frac{(K_1b)\psi_l^{(1)\prime}(K_1b)}{\psi_l^{(1)}(K_1b)} + \frac{1}{2}(K_1b)^2\frac{\mu_1}{\mu_2-\mu_1})}{CD_1}\right)$$

$$C_{22} = \frac{1}{\frac{\zeta_l^{(2)}(k_2b)}{\zeta_l^{(2)}(k_2a)}(F_2^L - \frac{(k_2b)\zeta_l^{(2)\prime}(k_2b)}{\zeta_l^{(2)}(k_2b)})}\left(F_2^L - 1 + \frac{\frac{1}{2}(K_2b)^2\frac{\mu_2}{\mu_2-\mu_1}(l(l+1) - \frac{(k_1b)h_l^{(1)\prime}(k_2b)}{h_l^{(1)}(k_2b)}(\frac{(K_1b)\psi_l^{(1)\prime}(K_1b)}{\psi_l^{(1)}(K_1b)} - \frac{1}{2}(K_1b)^2\frac{\mu_1}{\mu_2-\mu_1}))}{CD_1}\right)$$

$$Mult^V = \left(\frac{\zeta_l^{(1)}(k_2b)}{\zeta_l^{(1)}(k_2a)} - \frac{\zeta_l^{(2)}(k_2b)}{\zeta_l^{(2)}(k_2a)}\right)\left(\frac{\zeta_l^{(1)}(K_2b)}{\zeta_l^{(1)}(K_2a)} - \frac{\zeta_l^{(2)}(K_2b)}{\zeta_l^{(2)}(K_2a)}\right)\left(\frac{\mu 2 - \mu 1}{\mu 1}\right)^2\left(\frac{\mu 2 - \mu 3}{\mu 3}\right)^2 CD_1CD_2$$
$$\left(F_1^LF_1^T\frac{\zeta_l^{(2)}(k_2b)}{\zeta_l^{(2)}(k_2a)}\frac{\zeta_l^{(2)}(K_2b)}{\zeta_l^{(2)}(K_2a)}(F_2^L - \frac{(k_2b)\zeta_l^{(2)\prime}(k_2b)}{\zeta_l^{(2)}(k_2b)})(F_2^T - \frac{(K_2b)\zeta_l^{(2)\prime}(K_2b)}{\zeta_l^{(2)}(K_2b)})\right)$$

$$CD_1 = l(l+1)\left(2\left(1 - \frac{1}{2}(K_1b)^2\frac{\mu_1}{\mu_2-\mu_1}\right) - l(l+1) + \frac{(k_1b)h_l^{(1)\prime}(k_2b)}{h_l^{(1)}(k_2b)}\frac{(K_1b)\psi_l^{(1)\prime}(K_1b)}{\psi_l^{(1)}(K_1b)}\right)$$
$$-2\frac{(k_1b)h_l^{(1)\prime}(k_2b)}{h_l^{(1)}(k_2b)}\frac{(K_1b)\psi_l^{(1)\prime}(K_1b)}{\psi_l^{(1)}(K_1b)} + \frac{1}{2}(K_1b)^2\frac{\mu_1}{\mu_2-\mu_1}\left(2\frac{(k_1b)h_l^{(1)\prime}(k_2b)}{h_l^{(1)}(k_2b)} - \frac{1}{2}(K_1b)^2\frac{\mu_1}{\mu_2-\mu_1} + \frac{(K_1b)\psi_l^{(1)\prime}(K_1b)}{\psi_l^{(1)}(K_1b)}\right)$$

5.3.2 Dynamic stress concentrations

We may now write the explicit expressions for the dynamic stress concentrations due to an incident transverse displacement field. These can then be evaluated to determine the effect of an interface layer. We normalize the dynamic stress concentrations by calculating the incident stress field and computing σ^*. In this case, all stresses will be normalized to the incident shear stress. The radial stresses can be expressed as:

$$
\begin{aligned}
\sigma^*_{rr} &= \frac{\left|\sigma^L_{rr} + \sigma^V_{rr}\right|}{|\sigma^i_{xy}|} \\
\sigma^L_{rr} &= \sum_{n=0}^{\infty} \frac{2\mu_1}{(k_1 r)^2 r} \left(\frac{(K_1 r)^2}{2} - (l(l+1)+2) + 2(k_1 r)\frac{\zeta_l^{(1)\prime}(k_1 r)}{\zeta_l^{(1)}(k_1 r)} \right) \\
&\quad \frac{(2l+1)}{l(l+1)} \frac{\psi_l(K_1 b)}{\zeta_l^{(1)}(k_1 b)} \left(\frac{k_1}{K_1}\right)^2 P_l^{(1)} \cos\phi \left(\frac{\Delta_1^V}{\Delta_0}\right) i^{l+1} \zeta_l^{(1)}(k_1 r) \\
\sigma^V_{rr} &= \sum_{n=0}^{\infty} \frac{2\mu_1}{(K_1 r)^2 r} \left(l(l+1) \left(2 - (K_1 r) \frac{\zeta_l^{(1)\prime}(K_1 r)}{\zeta_l^{(1)}(K_1 r)} \right) \right) \\
&\quad \frac{(2l+1)}{l(l+1)} \frac{\psi_l(K_1 b)}{\zeta_l^{(1)}(K_1 b)} \left(\frac{\Delta_1^V}{\Delta_0}\right) P_l^{(1)} \cos\phi i^{l-1} \zeta_l^{(1)}(K_1 r)
\end{aligned} \tag{5.21}
$$

and the shear stress can be expressed as:

$$
\begin{aligned}
\sigma^*_{r\theta} &= \frac{\left|\sigma^L_{r\theta} + \sigma^V_{r\theta}\right|}{|\sigma^i_{xy}|} \\
\sigma^L_{r\theta} &= \sum_{n=0}^{\infty} \frac{2\mu_1}{(k_1 r)^2 r} \left(2 - (k_1 r) \frac{\zeta_l^{(1)\prime}(k_1 r)}{\zeta_l^{(1)}(k_1 r)} \right) \\
&\quad \frac{(2l+1)}{l(l+1)} \frac{\psi_l(K_1 b)}{\zeta_l^{(1)}(k_1 b)} \left(\frac{k_1}{K_1}\right)^2 P_l^{(1)} \cos\phi \left(\frac{\Delta_1^V}{\Delta_0}\right) i^{l+1} \zeta_l^{(1)}(k_1 r)
\end{aligned}
$$

$$\sigma_{r\theta}^{V} = \sum_{n=0}^{\infty} -\frac{2\mu_1}{(K_1 r)^2 r}\left(l(l+1) - \frac{(K_1 r)^2}{2} - (K_1 r)\frac{\zeta_l^{(1)\prime}(K_1 r)}{\zeta_l^{(1)}(K_1 r)}\right)$$

$$\frac{(2l+1)}{l(l+1)}\frac{\psi_l(K_1 b)}{\zeta_l^{(1)}(K_1 b)}\left(\frac{\Delta_1^V}{\Delta_0}\right) P_l^{(1)} \cos\phi i^{l-1}\zeta_l^{(1)}(K_1 r) \tag{5.22}$$

5.4 Incident uncoupled transverse wave

For an incident transverse wave of unit amplitude, traveling in the $\hat{z}$ direction and uncoupled to the longitudinal and second transverse field, we will have only scattered, refracted, and transmitted displacement fields of the same type generated. The incident π_S displacement field, scattered, refracted, and transmitted fields are given by the potentials:

$$(K_1 r)\pi_S^i = \sum_{l=1}^{\infty} i^l \left(\frac{2l+1}{l(l+1)}\right) \psi_l(K_1 r) P_l^{(1)}(\cos\theta)\sin\phi$$

$$(K_1 r)\pi_S^s = \sum_{l=1}^{\infty} B_{1l} i^l \zeta_l^{(1)}(K_1 r) P_l^{(1)}(\cos\theta)\sin\phi$$

$$(K_2 r)\pi_S^r = \sum_{l=1}^{\infty} B_{2l} i^l \zeta_l^{(1)}(K_2 r) P_l^{(1)}(\cos\theta)\sin\phi$$

$$(K_2 r)\tilde{\pi}_S^r = \sum_{l=1}^{\infty} \tilde{B}_{2l} i^l \zeta_l^{(2)}(K_2 r) P_l^{(1)}(\cos\theta)\sin\phi$$

$$(K_3 r)\pi_S^t = \sum_{l=1}^{\infty} B_{3l} i^l \psi_l(K_3 r) P_l^{(1)}(\cos\theta)\sin\phi \tag{5.23}$$

By applying the boundary conditions, we arrive at the four equations in four unknowns to solve. In matrix form we can write them as $Nx = c$. Filling the matrix N with the expressions for the displacements and normal surface tractions at the two interfaces gives the expressions for the elements of the matrix N:

$$N[1,1] = \frac{i}{K_1^2 b \sin\theta}$$

$$N[1,2] = -\frac{i}{K_1^2 b \sin\theta} \frac{\zeta_l^{(1)}(K_2 b)}{\zeta_l^{(1)}(K_2 a)}$$

$$N[1,3] = -\frac{i}{K_1^2 b \sin\theta} \frac{\zeta_l^{(2)}(K_2 b)}{\zeta_l^{(2)}(K_2 a)}$$

$$N[1,4] = 0$$

$$N[2,1] = \frac{i\mu_1}{(K_1 b)^2 \sin\theta} \left(2 - (K_1 b) \frac{\zeta_l^{(1)\prime}(K_1 b)}{\zeta_l^{(1)}(K_1 b)}\right)$$

$$N[2,2] = -\frac{i\mu_2}{(K_2 b)^2 \sin\theta} \left(2 - (K_2 b) \frac{\zeta_l^{(1)\prime}(K_2 b)}{\zeta_l^{(1)}(K_2 b)}\right) \frac{\zeta_l^{(1)}(K_2 b)}{\zeta_l^{(1)}(K_2 a)}$$

$$N[2,3] = -\frac{i\mu_2}{(K_2 b)^2 \sin\theta} \left(2 - (K_2 b) \frac{\zeta_l^{(2)\prime}(K_2 b)}{\zeta_l^{(2)}(K_2 b)}\right) \frac{\zeta_l^{(2)}(K_2 b)}{\zeta_l^{(2)}(K_2 a)}$$

$$N[2,4] = N[3,1] = 0$$

$$N[3,2] = -\frac{i\mu_2}{(K_2 a)^2 \sin\theta} \left(2 - (K_2 a) \frac{\zeta_l^{(1)\prime}(K_2 a)}{\zeta_l^{(1)}(K_2 a)}\right)$$

$$N[3,3] = -\frac{i\mu_2}{(K_2 b)^2 \sin\theta} \left(2 - (K_2 b) \frac{\zeta_l^{(2)\prime}(K_2 b)}{\zeta_l^{(2)}(K_2 b)}\right)$$

$$N[3,4] = \frac{i\mu_3}{(K_3 a)^2 \sin\theta} \left(2 - (K_3 a) \frac{\psi_l'(K_3 a)}{\psi_l(K_3 a)}\right)$$

$$N[4,1] = 0$$

$$N[4,2] = N[4,3] = -\frac{i}{K_2^2 a \sin\theta}$$

$$N[4,4] = \frac{i}{K_3^2 a \sin\theta} \tag{5.24}$$

In the above matrix, we have factored the radial functions as before in the scattered, refracted, and transmitted fields. We can now express the incident field in terms of the boundary conditions. The c column vector for an incident π_V transverse wave has, after factoring out the radial function common to all terms, the components:

$$c[1] = -\frac{i}{K_1^2 b \sin\theta}$$

$$c[2] = -\frac{i\mu_1}{(K_1 b)^2 \sin\theta}\left(2 - (K_1 b)\frac{\psi_l'(K_1 b)}{\psi_l(K_1 b)}\right)$$

$$c[3] = c[4] = 0 \tag{5.25}$$

The modal coefficients of the potentials can then be written in terms of the determinants as:

$$B_{1l} = -\frac{(2l+1)}{l(l+1)}\frac{\psi_l(K_1 b)}{\zeta_l^{(1)}(K_1 b)}\left(\frac{\Delta_1^V}{\Delta_0}\right)$$

$$B_{2l} = -\frac{(2l+1)}{l(l+1)}\frac{\psi_l(K_1 b)}{\zeta_l^{(1)}(K_2 b)}\left(\frac{K_2}{K_1}\right)^2\left(\frac{\Delta_2^V}{\Delta_0}\right)$$

$$\tilde{B}_{2l} = -\frac{(2l+1)}{l(l+1)}\frac{\psi_l(K_1 b)}{\zeta_l^{(2)}(K_2 b)}\left(\frac{K_2}{K_1}\right)^2\left(\frac{\tilde{\Delta}_2^V}{\Delta_0}\right)$$

$$B_{3l} = -\frac{(2l+1)}{l(l+1)}\frac{\psi_l(K_1 b)}{\psi_l(K_3 b)}\left(\frac{K_3}{K_1}\right)^2\left(\frac{\Delta_3^V}{\Delta_0}\right) \tag{5.26}$$

5.4.1 Scattered field and stress concentration

The scattered field from the spherical interface will be of the same type as incident in this case. The potential is given, after substitution of the expression for the modal coefficient, as:

$$(K_1 r)\pi_S^s = \sum_{l=1}^{\infty} -\frac{(2l+1)}{l(l+1)} \frac{\psi_l(K_1 b)}{\zeta_l^{(1)}(K_1 b)} \left(\frac{\Delta_1^V}{\Delta_0}\right) i^l \zeta_l^{(1)}(K_1 r) P_l^{(1)}(\cos\theta)\sin\phi \tag{5.27}$$

with the expressions for Δ_0 and Δ_1^V given as:

$$\Delta_1^S = \left(2 - \left(\frac{(K_2a)\zeta_l^{(2)\prime}(K_2a)}{\zeta_l^{(2)}(K_2a)} - F_1\frac{\zeta_l^{(2)}(K_2b)}{\zeta_l^{(2)}(K_2a)}\right) + \frac{\mu_3}{\mu_2}\left(\frac{(K_3a)\psi_l'(K_3a)}{\psi_l(K_3a)} - 2\right)\right)$$

$$\left(2 - F_2 + \frac{\mu_1}{\mu_2}\left(\frac{(K_1b)\psi_l'(K_1b)}{\psi_l(K_1b)} - 2\right)\right) + F_1\frac{\zeta_l^{(2)}(K_2b)}{\zeta_l^{(2)}(K_2a)}\left(F_2 - \frac{(K_2b)\zeta_l^{(2)\prime}(K_2b)}{\zeta_l^{(2)}(K_2b)}\right) \tag{5.28}$$

$$\Delta_0 = \left(2 - \left(\frac{(K_2a)\zeta_l^{(2)\prime}(K_2a)}{\zeta_l^{(2)}(K_2a)} - F_1\frac{\zeta_l^{(2)}(K_2b)}{\zeta_l^{(2)}(K_2a)}\right) + \frac{\mu_3}{\mu_2}\left(\frac{(K_3a)\psi_l'(K_3a)}{\psi_l(K_3a)} - 2\right)\right)$$

$$\left(2 - F_2 + \frac{\mu_1}{\mu_2}\left(\frac{(K_1b)\zeta_l^{(1)\prime}(K_1b)}{\zeta_l^{(1)}(K_1b)} - 2\right)\right) + F_1\frac{\zeta_l^{(2)}(K_2b)}{\zeta_l^{(2)}(K_2a)}\left(F_2 - \frac{(K_2b)\zeta_l^{(2)\prime}(K_2b)}{\zeta_l^{(2)}(K_2b)}\right) \tag{5.29}$$

$$F_1 = \frac{(K_2a)\left(\zeta_l^{(1)\prime}(K_2a)\zeta_l^{(2)}(K_2a) - \zeta_l^{(1)}(K_2a)\zeta_l^{(2)\prime}(K_2a)\right)}{\left(\zeta_l^{(1)}(K_2b)\zeta_l^{(2)}(K_2a) - \zeta_l^{(2)}(K_2b)\zeta_l^{(1)}(K_2a)\right)} \qquad F_2 = \frac{(K_2b)\left(\zeta_l^{(1)\prime}(K_2b)\zeta_l^{(2)}(K_2a) - \zeta_l^{(2)\prime}(K_2b)\zeta_l^{(1)}(K_2a)\right)}{\left(\zeta_l^{(1)}(K_2b)\zeta_l^{(2)}(K_2a) - \zeta_l^{(2)}(K_2b)\zeta_l^{(1)}(K_2a)\right)}$$

The expression for the scattered field can now be used to explicitly write the dynamic stress concentrations for shear stress as:

$$\sigma^*_{r\theta} = \frac{\left|\sigma^S_{r\theta}\right|}{\left|\sigma^i_{xy}\right|}$$

with the scattered shear stress given as:

$$\sigma^S_{r\theta} = \sum_{n=0}^{\infty} -\frac{i\mu_1}{(K_1 r)^2 \sin\theta}\left(2-(K_1 r)\frac{\zeta_l^{(1)\prime}(K_1 r)}{\zeta_l^{(1)}(K_1 r)}\right)$$

$$\frac{(2l+1)}{l(l+1)}\frac{\psi_l(K_1 b)}{\zeta_l^{(1)}(K_1 b)}\left(\frac{\Delta_1^V}{\Delta_0}\right) i^l \zeta_l^{(1)}(K_1 r) P_l^{(1)}(\cos\theta)\sin\phi \tag{5.30}$$

5.5 Applications to particle composites

There are many applications of this analysis of dynamical stress concentrations in particle reinforced composite materials. However, the lack of available comparisons in the literature demands a focus on the better understood behavior of the scattering cross sections essential to the verification of the analysis. The longitudinal scattering cross section is an integral of the energy flux over a spherical surface surrounding the sphere and normalized to the geometric cross section of the sphere (πb^2). This quantity is a measure of the amount of energy lost from the incident field due to scattering from the sphere and includes both the longitudinal and transverse scattered displacement fields. The longitudinal scattering cross section can be written as:

$$Q^L_{scatt} = \frac{4}{(k_1 b)^2}\sum_{l=0}^{\infty}(2l+1)\left\{\left|\frac{\Delta_1^L}{\Delta_0}\right|^2 + l(l+1)\frac{K_1}{k_1}\left|\frac{\Delta_2^L}{\Delta_0}\right|^2\right\} \tag{5.31}$$

Since this quantity involves both of the scattered field coefficients, it provides an excellent check of the algebraic inversion.

5.6 Special cases and verification

The primary source of comparison in the literature on this type of problem is through the scattering cross sections. This involves a sum of the coefficients produced by the scattering analysis and is performed here for the well tested analysis of Johnson and Truell [52]. The materials selected are figures 4-6 of the article. The first configuration tested in this analysis is beryllium with a thick shell of a beryllium like material embedded in polyethylene. The total scattering cross section shown in figure 5.2 is identical to figure 4 of the article. Also tested where a sphere of magnesium in stainless steel, and a sphere of stainless steel in magnesium shown in figures 5.3-5.4 and nearly identical to figures 4 & 5 of the article. A slight smoothing occurs at very high $k_1 b$ which is due to the presence, in this analysis, of a shell with physical properties nearly identical to the core. These results provide a good comparison and strong confirmation of the correctness of the solution. The cross sections are for a non-layered sphere and the analysis actually included a nearly identical layer of material as a shell. This capability to predict extreme cases is part of the advantage in formally solving the problem.

As a final case, we take the magnesium sphere and surround it with a shell of aluminum to predict the effects of an interface layer with the scattering cross section. We also let the ratio of inner to outer radius be large and small to study the range of behavior. In figure 5.5, the scattering cross sections for the two extremes of radius ratio are shown. The thin interface layer has little effect on the scattering cross section for low $k_1 b$, having the same behavior as the nonlayered magnesium. The thick interface layer, also in figure 5.5, gives significantly different behavior of the scattering cross section for all ranges of $k_1 b$.

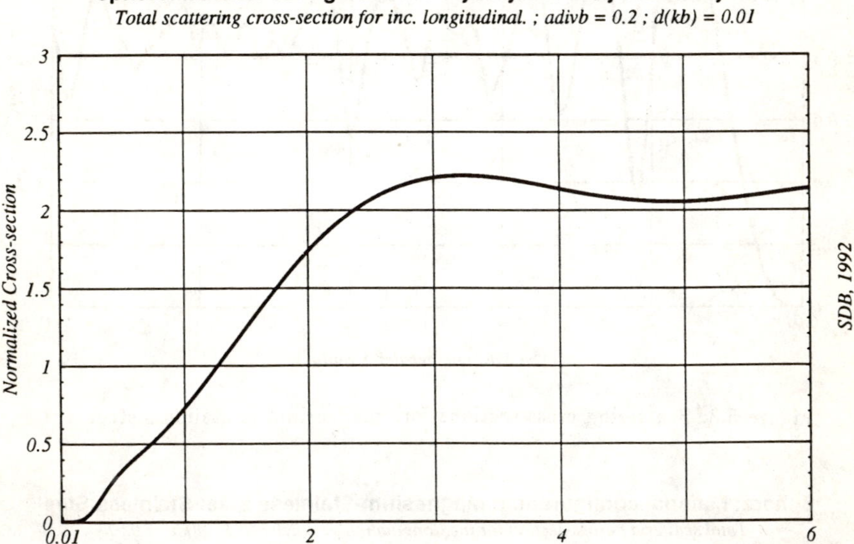

Figure 5.2: Scattering cross-sections for: beryllium in polyethylene.

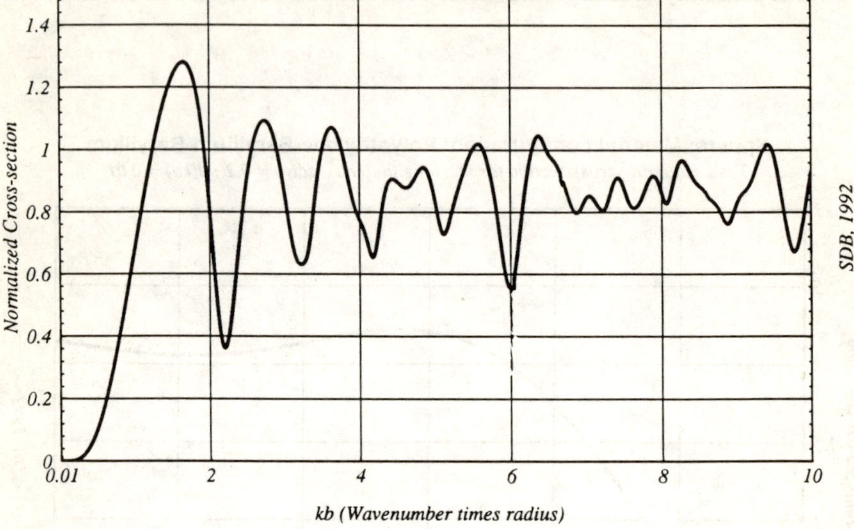

Figure 5.3: Scattering cross-sections for: magnesium in stainless steel.

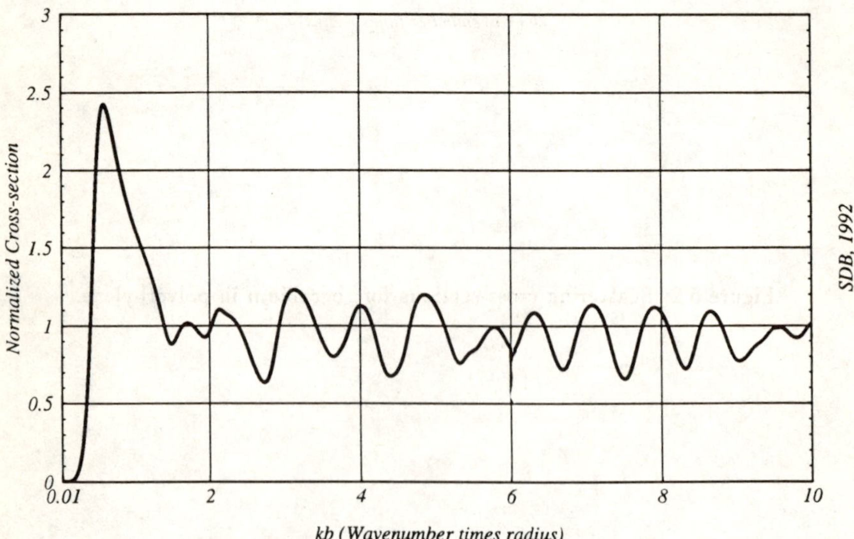

Figure 5.4: Scattering cross-sections for: stainless steel in magnesium.

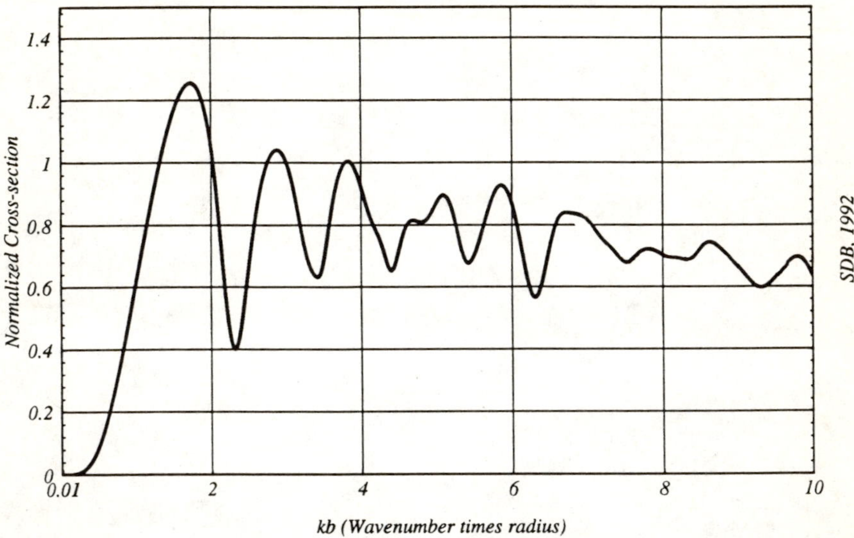

Figure 5.5: Limiting case: steel-exterior, Al-shell, Mg-core.

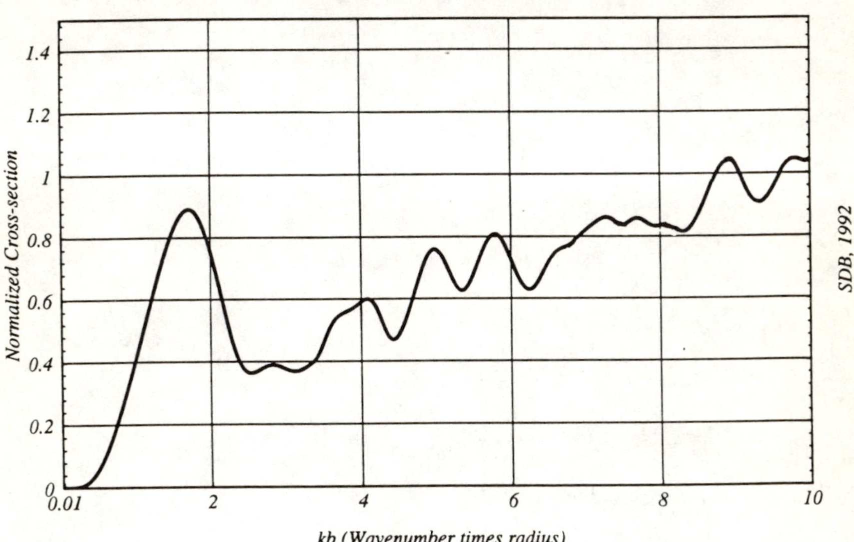

Figure 5.6: Limiting case: steel-exterior, Al-shell, Mg-core.

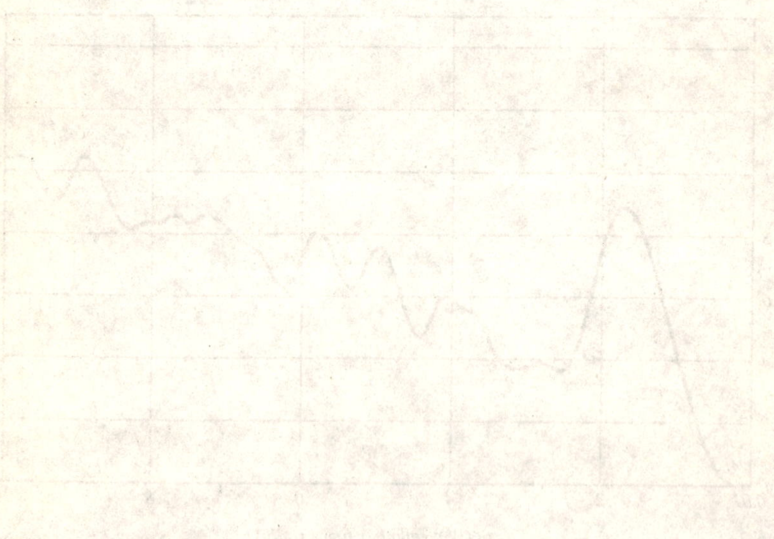

Chapter 6

SUMMARY OF RESULTS

The study of the effect of interface layers on the scattering of elastic waves has been presented for planar, cylindrical, and spherical interfaces. The general method of solution for the Mie scattering problem is presented and applied to the three cases with general expressions for the scattered displacement field amplitudes derived. In each case, the dynamic stress concentrations are studied as a function of the size of the scatterer and physical insight into the scattering problem is derived from the behavior of the dynamic stress concentrations or scattering cross sections.

The substantial theoretical framework for analyzing these types of problems was derived and presented. It remains, in future work, to extend this analysis to include additional interactions with the interface layers and to relate the dynamical problem more closely with the static problems of stress concentrations.

Among the future work intended, is to add thermal conductivity, as Epstein and Carhart did in their analysis [53] to predict the interaction of acoustic waves with fog. This is envisioned to provide a framework for predicting the effects of ultrasonic heating of tumors. In addition, more layers will be added to the problem to better model the multi-layered structures found in many man made and natural materials. This can be achieved in a way similar to the approach of Thomson [31]. While Thomson inverted the traction displacement relations, we intend to invert the displacement potential relations in a similar way and write a recursion relation for the multiple layers.

There are many other directions to take in this type of scattering analysis for elastic

waves. One area of great interest will be to develop the direct relation between this work and the work of Y.H. Pao and C.C. Mow [83] who, by using extreme values of material parameters, indicate that the elastic scattering analysis can be directly related to the acoustic scattering problem. This and many other related problem will provide many years of continued research, all with the foundations of the elastic displacement wave scattering by interface layers presented herein.

Appendix A

DETAILED DERIVATIONS FOR DYNAMICAL ELASTICITY

In formulating the equations of dynamical elasticity, we will begin by defining the relations for the deformation of a body under the action of an applied force. We will then relate the applied force to a stress-displacement field and introduce stress-strain relations and the assumptions of isotropy and homogeneity to arrive at the equilibrium (Navier) equations for a linear elastic, isotropic-homogeneous solid. The discussions that follow are similar to those of Eringen [84] and those of Fung [86] in their respective texts.

We begin by defining a material point P^0 of a body by a Euclidean coordinate system with coordinates a_i. A force is applied to the body displacing the point to $P^{0\prime}$ in the original coordinate system. Define a second Euclidean coordinate system where x_i gives the location of $P^{0\prime}$ in the second coordinate system as Q^0. Assuming that the deformation of the body lead to no discontinuities and all deformed points in the body can be similarly written as a unique point in the new coordinate system, we can then write the position of the deformed point in the second coordinate system as a mapping of the undeformed point in the original coordinate system

$$x_i = f_i(a_1, a_2, a_3, t) \tag{A.1}$$

Uniqueness and lack of discontinuities allow us to write an explicit inverse to the mapping function f as

$$a_i = f_i^{-1}(x_1, x_2, x_3, t) \tag{A.2}$$

Where this mapping function can be applied to any undeformed point on the body to arrive at the deformed point in the new coordinate system, or the inverse function can be applied to the deformed points in the second coordinate system to arrive at the original point in the first coordinate system. This is sometimes referred to as the axiom of continuity.

We now add an additional point P^1 to the original coordinate system in the neighborhood of P^0, letting $P^1 = P^0 + da$. These two points have the corresponding points Q^0 and Q^1 in the second (deformed) coordinate system, where $Q^1 = Q^0 + dx$. If we now join points P^0 and P^1 by a line segment whose length squared is

$$dl_0^2 = a_{ij} da^i da^j \tag{A.3}$$

we can similarly join points Q^0 and Q^1 in the deformed coordinate system with a line segment of length squared dl^2 given as

$$dl^2 = g_{ij} dx^i dx^j \tag{A.4}$$

The expressions a_{ij} and g_{ij} are the Euclidean metrics for the two coordinate systems. If we now take the difference between these two squared distances in the undeformed and deformed systems and expand the derivatives with the chain rule in the undeformed coordinate system, we arrive at

$$\begin{aligned} dl^2 - dl_0^2 &= g_{ij} dx^i dx^j - a_{ij} da^i da^j \\ dl^2 - dl_0^2 &= \left(g_{ij} - a_{mn} \frac{\partial a_m}{\partial x_i} \frac{\partial a_n}{\partial x_j} \right) dx^i dx^j \\ dl^2 - dl_0^2 &= e_{ij} \left(2 dx^i dx^j \right) \end{aligned} \tag{A.5}$$

where e_{ij} is the Almansi strain tensor, referred to as the Euclidean strain tensor or Cauchy strain tensor for infinitesimal displacements. We now introduce the displacement vector $u_i = x_i - a_i$, then we can write the strain tensor in terms of displacements. In cartesian coordinates, they take on the form

$$e_{ij} = \frac{1}{2}\left(\frac{\partial u_i}{\partial x_j} + \frac{\partial u_j}{\partial x_i} - \frac{\partial u_m}{\partial x_i}\frac{\partial u_m}{\partial x_j}\right) \tag{A.6}$$

Having derived the strain tensor, our task is to relate the forces that have caused the deformations to the displacement vector and strain tensor. We begin by applying Newton's laws of motion for particles, extended by Euler to continuous media, to the body under an applied force. The resultant of forces acting on an arbitrary volume is equal to the rate of change of linear momentum

$$\frac{D}{Dt}M_i = F_i \tag{A.7}$$

where the linear momentum, M_i is given by

$$M_i = \int_{Vol} \rho v_i d^3x \tag{A.8}$$

with v_i the velocity and ρ density. The applied forces F_i, are external forces applied to the surface of the volume through the normal surface tractions T_i, and the internal body forces B_i. The total force can then be written as

$$F_i = \int_{Surface} T_i d^2x \quad + \int_{Vol} B_i d^3x \tag{A.9}$$

The derivative of the linear momentum in equation A.7 can be rewritten upon substitution of A.8 as

$$\frac{D}{Dt}M_i = \int_{Vol}\left(\frac{\partial(\rho v_i)}{\partial t} + \frac{\partial}{\partial x_j}(\rho v_i v_j)\right) d^3x \tag{A.10}$$

which can be further reduced, given that no mass is lost in the deformation, to

$$\frac{D}{Dt}M_i = \int_{Vol} \rho \frac{Dv_i}{Dt} d^3x \tag{A.11}$$

We can now substitute A.11 into the equilibrium equation with the normal surface traction replaced by the contraction of the stress tensor with the unit surface normal. The equilibrium equation can then be written as a single volume integral using the divergence theorem as

$$\int_{Vol}\left(\frac{\partial \sigma_{ij}}{\partial x_j}+B_i\right)d^3x=\int_{Vol}\rho\frac{Dv_i}{Dt}d^3x \tag{A.12}$$

Allowing for equilibrium of an arbitrary volume we have the equations of motion of Euler

$$\left(\frac{\partial \sigma_{ij}}{\partial x_j}+B_i\right)=\rho\frac{Dv_i}{Dt} \tag{A.13}$$

The velocity v_i is then expanded in terms of the displacement vector as

$$v_i=v_j\frac{\partial u_i}{\partial x_j}+\frac{\partial u_i}{\partial t} \tag{A.14}$$

and the stress tensor, if taken as isotropic and homogeneous, can be written in terms of the strain tensor as

$$\sigma_{ij}=\lambda e_{mm}\delta_{ij}+2\mu e_{ij} \tag{A.15}$$

Before substituting into the equations of motion we make two linearizations. The first to the strain tensor (A.6), taking only the first two terms, and the second to the velocity (A.14) by taking only the last term to give

$$e_{ij}=\frac{1}{2}\left(\frac{\partial u_i}{\partial x_j}+\frac{\partial u_j}{\partial x_i}\right) \tag{A.16}$$

and

$$v_i=\frac{\partial u_i}{\partial t} \tag{A.17}$$

Now substituting the above expressions into A.13 and neglecting the internal body forces gives Navier's equations of motion, rewritten in vector form for convenience, which is our starting point in the analysis of problems in dynamical elasticity

$$\rho\frac{\partial^2\vec{u}}{\partial t^2}-\mu\nabla^2\vec{u}-(\mu+\lambda)\nabla(\nabla\cdot\vec{u})=0 \tag{A.18}$$

Appendix B

MATHEMATICAL DISCUSSIONS

B.1 Separation and solution of the vector Helmholtz equation

In solving the equations of dynamical elasticity, we note that the equations of motion can be written, using the Helmholtz vector field decomposition, as two vector Helmholtz equations:

$$\left(\nabla^2 + k^2\right)\vec{u}_L = 0 \qquad \left(\nabla^2 + K^2\right)\vec{u}_T = 0 \tag{B.1}$$

where

$$\nabla \times \vec{u}_L = 0 \qquad \nabla \cdot \vec{u}_T = 0 \tag{B.2}$$

The longitudinal field $\vec{u}_L$, is written as the gradient of a potential function, giving the scalar Helmholtz equation for the longitudinal vector field. We can express these as

$$\vec{u}_L = \nabla\phi \qquad \left(\nabla^2 + k^2\right)\nabla\phi = \nabla\left(\nabla^2 + k^2\right)\phi = 0 \tag{B.3}$$

The solutions to the scalar Helmholtz equation are known for many different coordinate systems and can be easily written for the coordinate systems under study. No difficulty arrises with the longitudinal field, since the gradient transposes with the Helmholtz operator $(\nabla^2 + k^2)$. It is the transverse field that introduces difficulties in finding solutions to the vector Helmholtz equation. Substituting the curl into the vector Helmholtz equation gives:

$$\vec{u}_T = \nabla \times \vec{A} \qquad \left(\nabla^2 + K^2\right)\nabla \times \vec{A} = 0 \tag{B.4}$$

To find solutions for $\vec{A}$, we follow the approach of Morse & Feshbach, introducing a base vector $\hat{a}_1$, scalar function $w(x_1)$, and potential field ψ which satisfies the scalar Helmholtz equation

$$\left(\nabla^2 + K^2\right) \nabla \times (\hat{a}_1 w(x_1)\psi) = 0 \qquad \left(\nabla^2 + K^2\right) \psi = 0 \tag{B.5}$$

This leads to specific requirement on the values of $\hat{a}_1$ and $w(x_1)$ that give a scalar Helmholtz equation for ψ. As discussed in Morse & Feshbach, section 13.1 [48], these requirements are:

(1) One of the scale factors of the coordinate system must be unity.

(2) The ratio of the other two scale factors is independent of the coordinate corresponding to the unity scale factor.

Of the eleven coordinate systems that allow a separation of variables solution to the scalar Helmholtz equation, only six satisfy these requirements. A set of $\hat{a}_1$ and $w(x_1)$ can be developed from these requirements. In cartesian coordinates:

$$\hat{a}_1 = \hat{x},\ \hat{y},\ or\ \hat{z} \qquad w(x_1) = 1 \tag{B.6}$$

and in circular, elliptic, and parabolic cylindrical coordinates:

$$\hat{a}_1 = \hat{z} \qquad w(x_1) = 1 \tag{B.7}$$

The values for spherical and conical coordinates are:

$$\hat{a}_1 = \hat{r} \qquad w(x_1) = r \tag{B.8}$$

We began with a vector partial differential equation with three equations and three unknown scalar functions to solve for. The introduction of the first two potentials provides two scalar functions with the third scalar function either zero, in highly symmetric problems, or necessary to solving the problem in all other cases. A third potential can be introduced to complete the system in exactly the same manner as the first transverse field was introduced as

$$\left(\nabla^2 + K^2\right) \nabla \times \nabla \times (\hat{a}_1 w(x_1)\chi) = 0 \qquad \left(\nabla^2 + K^2\right) \chi = 0 \tag{B.9}$$

The same six coordinate systems allow a reduction to the scalar Helmholtz equation, with the same expressions for $\hat{a}_1$ and $w(x_1)$ previously given. When a solution is attempted in one of the remaining five coordinate systems: toroidal, oblate spheroidal, prolate spheroidal, paraboloidal, and parabolic coordinates, numerical solutions to the boundary value problems are then the only known alternative. In principle, all three components of the vector field could be decoupled and solved if we had the solutions to the three resulting sixth order ordinary differential equations.

We will focus on the solutions in planar, cylindrical, and spherical coordinates. The separation of variables approach to solving the scalar Helmholtz equation in these coordinate systems is discussed below.

B.1.1 Solutions in cartesian coordinates

In cartesian coordinates, the scalar Helmholtz equation

$$\left(\nabla^2 + k^2\right) \pi = 0 \tag{B.10}$$

can be written as

$$\frac{\partial^2 \pi}{\partial x} + \frac{\partial^2 \pi}{\partial y} + \frac{\partial^2 \pi}{\partial z} + k^2 \pi = 0 \tag{B.11}$$

Let $\pi = X(x)Y(y)Z(z)$, and dividing by $X(x)Y(y)Z(z)$ giving

$$\frac{X''}{X} + \frac{Y''}{Y} + \frac{Z''}{Z} + k^2 = 0 \tag{B.12}$$

The first three terms depend only on x, y, z respectively. They must therefore be all equal to constants for there to be a solution to the equation. If we choose those constants with some forethought as to the proper solutions for the problem under study, we arrive at the following definitions:

$$\frac{X''}{X} = -\alpha^2 \qquad \frac{Y''}{Y} = -\beta^2 \qquad \frac{Z''}{Z} = -\gamma^2 \tag{B.13}$$

Then we can easily write the solutions to these second order differential equations as:

$$X(x) = X_0 e^{\pm i\alpha x} \qquad Y(y) = Y_0 e^{\pm i\beta y} \qquad Z(z) = Z_0 e^{\pm i\gamma z} \tag{B.14}$$

with the following constraint on the values of α, β, γ.

$$\alpha^2 + \beta^2 + \gamma^2 = k^2 \tag{B.15}$$

The general solution for π can then be written as

$$\pi = \left\{X_0 e^{\pm i\alpha x}\right\}\left\{Y_0 e^{\pm i\beta y}\right\}\left\{Z_0 e^{\pm i\gamma z}\right\} \tag{B.16}$$

where the use of the $\{\ \}$ indicates that the full solution is a linear combination of these expressions. The relations for (α, β, γ) used in this analysis are for no evanescent modes, expressed as

$$\alpha = \vec{k}\cdot\hat{x} \quad \beta = \vec{k}\cdot\hat{y} \quad \gamma = \vec{k}\cdot\hat{z} \tag{B.17}$$

where $\vec{k}$ is the wavenumber multiplied by the normalized vector pointing in the direction of propagation of the wave.

B.1.2 Solutions in cylindrical coordinates

In cylindrical coordinates, the scalar Helmholtz equation

$$\left(\nabla^2 + k^2\right)\pi = 0 \tag{B.18}$$

can be written as

$$\left(\frac{1}{r}\frac{\partial}{\partial r}\left(r\frac{\partial}{\partial r}\right)+\frac{1}{r^2}\frac{\partial^2}{\partial\theta^2}+\frac{\partial^2}{\partial z^2}+k^2\right)\pi=0 \tag{B.19}$$

Let $\pi = R(r)\Theta(\theta)Z(z)$, and dividing by $R(r)\Theta(\theta)Z(z)$ gives

$$\frac{1}{R}\frac{d^2R}{dr^2}+\frac{1}{Rr}\frac{dR}{dr}+\frac{1}{r^2\Theta}\frac{d^2\Theta}{d\theta^2}+\frac{1}{Z}\frac{d^2Z}{dz^2}+k^2=0 \tag{B.20}$$

We note that the fourth term is a function of z only. Using γ as the separation constant and solving for $Z(z)$ gives:

$$\frac{1}{Z}\frac{d^2Z}{dz^2}=-\gamma^2 \qquad Z(z)=Z_0e^{\pm i\gamma z} \tag{B.21}$$

If we now insert γ into the previous equation and multiply by r^2, we can similarly separate out the θ and r dependence and solve for the remaining Θ and R functions

$$\frac{r^2}{R}\frac{d^2R}{dr^2}+\frac{r}{R}\frac{dR}{dr}+\left(k^2-\gamma^2\right)r^2+\frac{1}{\Theta}\frac{d^2\Theta}{d\theta^2}=0 \tag{B.22}$$

and

$$\frac{1}{\Theta}\frac{d^2\Theta}{d\theta^2}=-n^2 \qquad \Theta(\theta)=\Theta_0e^{\pm in\theta} \tag{B.23}$$

The remaining radial equation is then divided by r^2 giving Bessel's equation with Bessel and Neumann function solutions:

$$\frac{1}{R}\frac{d^2R}{dr^2}+\frac{1}{rR}\frac{dR}{dr}+\left(\left(k^2-\gamma^2\right)-\frac{n^2}{r^2}\right)=0 \tag{B.24}$$

$$R(r)=R_0\left\{\begin{array}{c}J_n(\sqrt{k^2-\gamma^2}r)\\ Y_n(\sqrt{k^2-\gamma^2}r)\end{array}\right\} \tag{B.25}$$

The general solution for π can then be written as:

$$\pi=\left\{\begin{array}{c}J_n(\sqrt{k^2-\gamma^2}r)\\ Y_n(\sqrt{k^2-\gamma^2}r)\end{array}\right\}\left\{e^{\pm in\theta}\right\}\left\{e^{\pm i\gamma z}\right\} \tag{B.26}$$

where the use of the { } indicates that the full solution is a linear combination of these expressions. The relation used for the argument to the radial functions is, for no evanescent modes, given by

$$\gamma = \vec{k} \cdot \hat{z} \tag{B.27}$$

where $\vec{k}$ is the wavenumber multiplied by the normalized vector pointing in the direction of propagation of the wave as in cartesian coordinates.

B.1.3 Solutions in spherical coordinates

In spherical coordinates, the scalar Helmholtz equation

$$\left(\nabla^2 + k^2\right)\pi = 0 \tag{B.28}$$

can be written as

$$\left(\frac{1}{r^2}\frac{\partial}{\partial r}\left(r^2\frac{\partial}{\partial r}\right) + \frac{1}{r^2 \sin\theta}\frac{\partial}{\partial\theta}\left(\sin\theta\frac{\partial}{\partial\theta}\right) + \frac{1}{r^2\sin^2\theta}\frac{\partial^2}{\partial\phi^2} + k^2\right)\pi = 0 \tag{B.29}$$

Let $\pi = R(r)\Theta(\theta)\Phi(\phi)$, and dividing through by $R(r)\Theta(\theta)\Phi(\phi)$ and multiplying through by $r^2\sin\theta$ gives

$$\frac{\sin\theta}{R}\frac{d}{dr}\left(r^2\frac{dR}{dr}\right) + \frac{\sin\theta}{\Theta}\frac{d}{d\theta}\left(\sin\theta\frac{d\Theta}{d\theta}\right) + \frac{1}{\Phi}\frac{d^2\Phi}{d\phi^2} + r^2\sin^2\theta k^2 = 0 \tag{B.30}$$

We note that the third term is a function of ϕ only. Using a separation constant of $-m^2$ and solving the $\Phi(\phi)$ we arrive at:

$$\frac{1}{\Phi}\frac{d^2\Phi}{d\phi^2} = -m^2, \qquad \Phi(\phi) = \Phi_0 e^{\pm im\phi} \tag{B.31}$$

If we now insert $-m^2$ back into the original separation equation and multiply by r^2, we can similarly separate out the θ and r dependence. Let's first tackle the Θ equation which leads the associated Legendre functions. The radial and angular equations are:

$$\frac{1}{R}\frac{d}{dr}\left(r^2\frac{dR}{dr}\right)+\frac{1}{\sin\theta\Theta}\frac{d}{d\theta}\left(\sin\theta\frac{d\Theta}{d\theta}\right)-\frac{m^2}{\sin^2\theta}+k^2=0 \tag{B.32}$$

and

$$\frac{1}{\sin\theta\Theta}\frac{d}{d\theta}\left(\sin\theta\frac{d\Theta}{d\theta}\right)-\frac{m^2}{\sin^2\theta}=-l(l+1) \tag{B.33}$$

If we let $x = \cos\theta$, and $dx = -\sin\theta d\theta$, we can rewrite the previous equation in the appropriate form for the associated Legendre functions given by

$$\left(1-x^2\right)\frac{d^2\Theta}{dx^2}-2x\frac{d\Theta}{dx}+l(l+1)\Theta-\frac{m^2}{1-x^2}\Theta=0 \tag{B.34}$$

$$\Theta=\left\{\begin{array}{c} P_l^m(\cos\theta) \\ Q_l^m(\cos\theta) \end{array}\right\} \tag{B.35}$$

The radial equation, after substituting in for $-l(l+1)$, can be reduced to

$$\frac{1}{r^2R}\frac{d}{dr}\left(r^2\frac{dR}{dr}\right)-\frac{l(l+1)}{r^2}+k^2=0 \tag{B.36}$$

If we let $R=\frac{\xi}{\sqrt{z}}$ and $z = kr$, the radial equation can then be written in the familiar form

$$\frac{d^2\xi}{dz^2}+\frac{2}{z}\frac{d\xi}{dz}+\left(1-\frac{l(l+1)}{z^2}\right)\xi=0 \tag{B.37}$$

with solutions in terms of cylindrical Bessel functions

$$\xi=\left\{\begin{array}{c} \sqrt{\frac{\pi}{2z}}J_{l+\frac{1}{2}}(z) \\ \sqrt{\frac{\pi}{2z}}Y_{l+\frac{1}{2}}(z) \end{array}\right\} \tag{B.38}$$

or writing as spherical Bessel functions

$$\xi=\left\{\begin{array}{c} j_l(kr) \\ y_l(kr) \end{array}\right\} \tag{B.39}$$

The general solution for π can then be written as

$$\pi=\left\{\begin{array}{c} j_l(kr) \\ y_l(kr) \end{array}\right\}\left\{\begin{array}{c} P_l^m(\cos\theta) \\ Q_l^m(\cos\theta) \end{array}\right\}\left\{e^{\pm im\phi}\right\} \tag{B.40}$$

We also note the relations between the Ricatti Bessel and Hankel functions, since we will primarily use the Ricatti functions in the eigenfunction expansions due to their natural occurrence in the incident field decomposition into spherical coordinates. The relations are:

$$\psi_l(kr) = (kr)j_l(kr) = \sqrt{\frac{\pi kr}{2}} J_{l+1/2}(kr)$$

$$\chi_l(kr) = (kr)y_l(kr) = \sqrt{\frac{\pi kr}{2}} Y_{l+1/2}(kr)$$

$$\zeta_l^{(1,2)}(kr) = (kr)h_l^{(1,2)}(kr) = \sqrt{\frac{\pi kr}{2}} H_{l+1/2}^{(1,2)}(kr)$$

$$\zeta_l^{(1,2)}(kr) = \psi_l(kr) \pm i\chi_l(kr)$$

Bibliography

[1] A. Clebsch: Crelle's Journal **61** 195 (1863).

[2] L. Lorenz: Videnskab. Skrifter. **6** (1890).

[3] J.W.S. Rayleigh: Phil. Mag. **XLI**, 107, 187 & 447 (1871).

[4] H. Lamb: Proc. Lond. Math. Soc. **XXXII**, 11, 120 (1900).

[5] C.J.T. Sewel: Phil. Trans. A. **CCX**, 239 (1910).

[6] K. F. Herzfeld: Phil. Mag **9**, 741 (1930).

[7] V. C. Anderson: JASA **22**, #4, 426 (1950).

[8] J. J. Faran: JASA **23**, #4, 405 (1951).

[9] C. F. Ying and R. Truell: J. Appl. Phys. **27**, 1087 (1956).

[10] N. G. Einspruch, E. J. Witterholt and R. Truell: Jour. Appl. Phys. **31**, 806 (1960).

[11] N. G. Einspruch and R. Truell: JASA **32**, # 2 (1960).

[12] L. Knopoff: Geophys. **XXIV**, #1, 30 (1959).

[13] L. Knopoff: Geophys. **XXIV**, #2, 209 (1959).

[14] D. W. Kraft and M. C. Franzblau: J. Appl. Phys. **42**, #8, 3019 (1971).

[15] M. Nagase: J. Phys. Soc. Japan **11**, #3, 279 (1956).

[16] Y. Iwashimizu: J. Sound Vib. **21**, #4, 463 (1972).

[17] D. L. Jain and R. P. Kanwal: J. Sound Vib. **57**, #2, 171 (1978).

[18] V. K. Varadan and V. V. Varadan: Acoustic, Electromagnetic and Elastic Wave Scattering–Focus on the T-Matrix Approach (Pergamon Press, New York, 1980).

[19] L. Flax, G. C. Gaunaurd and H. Uberall: *Theory of Resonance Scattering*, in Physical Acoustics, Vol. XV (Academic Press, New York, 1981).

[20] L. Flax, L. R. Dragonette and H. Uberall: JASA **63**, #3, 732 (1978).

[21] D. Brill, G. Gaunaurd and H. Uberall: JASA **67**, #2, 414 (1980).

[22] L. Flax and H. Uberall: JASA **67**, #5, 1432 (1980).

[23] D. Brill and G. Gaunaurd: JASA **81**, #1, 1 (1987).

[24] J. E. Gubernatis, E. Domany and J. A. Krumhansl: J. Appl. Phys. **48**, #7, 2804 (1977).

[25] J. E. Gubernatis, E. Domany and J. A. Krumhansl: J. Appl. Phys. **48**, #7, 2812 (1977).

[26] J. E. Gubernatis: J. Appl. Phys. **50**, #6, 4046 (1979).

[27] W. Kohn and J. R. Rice: J. Appl. Phys. **50**, #5, 3346 (1979).

[28] G. G. Stokes: Trans. Camb. Phil. Soc. **IX**, 1 (1849).

[29] J.W. Nicholson: Phil. Mag **11**, 193 (1906).

[30] R. B. Lindsay: JASA **11**, 178 (1939).

[31] W. T. Thomson: J. Appl. Phys. **21**, 89 (1950).

[32] W. M. Ewing, W. S. Jardetsky and F. Press: Elastic Waves in Layered Media (McGraw-Hill, New York, 1957).

[33] L. M. Brekhovsleikh: Waves in Layered Media (Academic Press, New York, 1960).

[34] J. Lekner: Theory of Reflections (Martinus Nijhoff, Dordrecht, 1987).

[35] A. Gerard: Int. J. Eng. Sci. **17**, 313 (1979).

[36] R. R. Goodman and R. Stern: JASA **34**, 338 (1962).

[37] S. Machlup: JASA **24**, 290 (1952).

[38] R. Hickling: JASA **36**, 1124 (1964).

[39] M. C. Junger: JASA **24**, 366 (1952).

[40] T. K. Stanton: JASA **80**, 1619 (1990).

[41] G. Mie: Ann. Phys. **25**, 377 (1908).

[42] M.K. Hinders: Attenuation of Elastic Waves Due to Scattering from Spherical Cavities and Elastic Inclusions (Boston University Dissertation, Boston, 1990).

[43] M.K. Hinders: Phys. Rev. A. **43**, # 10, 5628, (1991).

[44] M.K. Hinders: Il. Nuovo Cimento, **106B**, # 7, 799, (1991).

[45] M.K. Hinders et al: Dynamic Stress Concentrations in Particle- and Fiber- Reinforced Composite Materials, in "Topics in Composite Materials and Structures," V. Birman and A. Nagar, eds. ASME AMD-Vol. 133, pp. 1 -11, (1992).

[46] M.K. Hinders: Il. Nuovo Cimento, **108B**, in press, (1993).

[47] S.D. Bogan and M.K. Hinders: Il. Nuovo Cimento, **107B**, #10, 1215, (1992).

[48] P. M. Morse and H. Feshbach: Methods of Theoretical Physics (McGraw-Hill, New York, 1953).

[49] M. Born and E. Wolf: Principles of Optics (Pergamon Press, New York, 1926).

[50] A. Yildiz: Il. Nuovo Cimento **XXX**, #5, 1182 (1963).

[51] P. Olsson, S.K.Datta, A. Bostrom: Trans. of the ASME, **57**, 672 (1990).

[52] G. Johnson and R. Truell: Jour. Appl. Phys. **36**, #11 3466 (1965).

[53] P. Epstein and R. Carhart: JASA **25**, # 3 (1953).

[54] A. Gerard: JASA **73**, # 1 (1983).

[55] S.K. Datta: JASA **61**, # 6 (1977).

[56] S.K. Datta, H.M Ledbetter, Y. Shindo, A.H. Shah: Wave Motion, **10**, 171 (1988).

[57] L.H. Huang and A.T. Chwang: Wave Motion, **12**, 401 (1990).

[58] F.E. Fox: JASA **12**, 147 (1940).

[59] A. Akay: JASA **89**, # 4 part I (1991).

[60] T. Hasegawa, K. Yosioka: JASA **46**, # 5 part II (1969).

[61] K. Kiriaki and D. Polyzos: Int. J. Eng. Sci. **26**, # 11 1143 (1988).

[62] R. McBride and D. Kraft: Jour. Appl. Phys. **43**, #12 (1972).

[63] Y.H. Pao and C.C. Mow: Jour. Appl. Phys. **34**, #3 493 (1963).

[64] W.T. Thomson: Jour. Appl. Phys. **21**, (1950).

[65] D.H. Hageman and V.L. Glass: Jour. Appl. Phys. **44**, # 9 (1973).

[66] H. Garnet and J. Crouzet-Pascal: Trans. of ASME, Jour. of Appl. Mech. 521 (1966).

[67] B. Peterson, V.K. Varadan, V.V. Varadan: JASA **73**, # 5 (1983).

[68] R.M. White: JASA **30**, # 8 771 (1958).

[69] W.H. Lin and A.C. Raptis: JASA **73**, # 3 736 (1983).

[70] F. Leon et al: JASA **91**, # 3 (1992).

[71] A.H. Nayfeh: JASA **89**, # 4 Part I (1991).

[72] X. Fan and K. Hynynon: JASA **91**, # 3 (1992).

[73] J.R. Yeh: Int. J. of Solid Structures **29**, # 20 2493 (1992).

[74] C. Sve: Trans. of ASME, J. Appl. Mech. (1971).

[75] N.A. Logan: Proc. of IEEE, 773 (1965).

[76] A.K. Chan, R.A. Sigelmann, A.W. Guy: IEEE trans. on Biomed. Eng., **BME-21** # 4 280 (1974).

[77] A.I. Beltzer: Acoustic of solids, Springer-Verlag, Chp V (1988).

[78] C.C. Mow and L.J. Mente: Trans. of ASME, J. of App. Mech., (1963)

[79] N.A. Haskel: Bull. Seis. Soc. Amer., **43** # 1 (1953)

[80] A. Norris and Y. Yang: Trans. of ASME, J. of App. Mech., **58**, 404 (1991)

[81] Lord Rayleigh: Theory of Sound, Macmillan (1878), 1894 edition reprinted by Dover.

[82] A.E.H. Love: A Treatise on the Mathematical Theory of Elasticity, Cambridge University Press (1927), reprinted by Dover.

[83] Y.H. Pao and C.C. Mow: Diffraction of Elastic Waves and Dynamic Stress Concentrations, The Rand Corp. (1973).

[84] A.C. Eringen: Mechanics of Continua, John Wiley & Sons, Inc. (1967).

[85] W.F. Smith: Principles of Material Science, McGraw Hill, pg769 (1990).

[86] Y.C. Fung: Foundations of solid mechanics, Prentice Hall (1965).

[87] R. H. Rand: Computer Algebra in Applied Mathematics (Research Notes in Mathematics, 94, Pitman, Boston, 1984).

Printing: Druckhaus Beltz, Hemsbach
Binding: Buchbinderei Schäffer, Grünstadt

Lecture Notes in Physics

For information about Vols. 1–388
please contact your bookseller or Springer-Verlag

Vol. 389: J. C. Miller, R. F. Haglund (Eds.), Laser Ablation-Mechanisms and Applications. Proceedings. IX, 362 pages, 1991.

Vol. 390: J. Heidmann, M. J. Klein (Eds.), Bioastronomy - The Search for Extraterrestrial Life. Proceedings, 1990. XVII, 413 pages. 1991.

Vol. 391: A. Zdziarski, M. Sikora (Eds.), Ralativistic Hadrons in Cosmic Compact Objects. Proceedings, 1990. XII, 182 pages. 1991.

Vol. 392: J.-D. Fournier, P.-L. Sulem (Eds.), Large-Scale Structures in Nonlinear Physics. Proceedings. VIII, 353 pages. 1991.

Vol. 393: M. Remoissenet, M.Peyrard (Eds.), Nonlinear Coherent Structures in Physics and Biology. Proceedings. XII, 398 pages. 1991.

Vol. 394: M. R. J. Hoch, R. H. Lemmer (Eds.), Low Temperature Physics. Proceedings. X,374 pages. 1991.

Vol. 395: H. E. Trease, M. J. Fritts, W. P. Crowley (Eds.), Advances in the Free-Lagrange Method. Proceedings, 1990. XI, 327 pages. 1991.

Vol. 396: H. Mitter, H. Gausterer (Eds.), Recent Aspects of Quantum Fields. Proceedings. XIII, 332 pages. 1991.

Vol. 398: T. M. M. Verheggen (Ed.), Numerical Methods for the Simulation of Multi-Phase and Complex Flow. Proceedings, 1990. VI, 153 pages. 1992.

Vol. 399: Z. Švestka, B. V. Jackson, M. E. Machedo (Eds.), Eruptive Solar Flares. Proceedings, 1991. XIV, 409 pages. 1992.

Vol. 400: M. Dienes, M. Month, S. Turner (Eds.), Frontiers of Particle Beams: Intensity Limitations. Proceedings, 1990. IX, 610 pages. 1992.

Vol. 401: U. Heber, C. S. Jeffery (Eds.), The Atmospheres of Early-Type Stars. Proceedings, 1991. XIX, 450 pages. 1992.

Vol. 402: L. Boi, D. Flament, J.-M. Salanskis (Eds.), 1830-1930: A Century of Geometry. VIII, 304 pages. 1992.

Vol. 403: E. Balslev (Ed.), Schrödinger Operators. Proceedings, 1991. VIII, 264 pages. 1992.

Vol. 404: R. Schmidt, H. O. Lutz, R. Dreizler (Eds.), Nuclear Physics Concepts in the Study of Atomic Cluster Physics. Proceedings, 1991. XVIII, 363 pages. 1992.

Vol. 405: W. Hollik, R. Rückl, J. Wess (Eds.), Phenomenological Aspects of Supersymmetry. VII, 329 pages. 1992.

Vol. 406: R. Kayser, T. Schramm, L. Nieser (Eds.), Gravitational Lenses. Proceedings, 1991. XXII, 399 pages. 1992.

Vol. 407: P. L. Smith, W. L. Wiese (Eds.), Atomic and Molecular Data for Space Astronomy. VII, 158 pages. 1992.

Vol. 408: V. J. Martínez, M. Portilla, D. Sàez (Eds.), New Insights into the Universe. Proceedings, 1991. XI, 298 pages. 1992.

Vol. 409: H. Gausterer, C. B. Lang (Eds.), Computational Methods in Field Theory. Proceedings, 1992. XII, 274 pages. 1992.

Vol. 410: J. Ehlers, G. Schäfer (Eds.), Relativistic Gravity Research. Proceedings, VIII, 409 pages. 1992.

Vol. 411: W. Dieter Heiss (Ed.), Chaos and Quantum Chaos. Proceedings, XIV, 330 pages. 1992.

Vol. 412: A. W. Clegg, G. E. Nedoluha (Eds.), Astrophysical Masers. Proceedings, 1992. XX, 480 pages. 1993.

Vol. 413: Aa. Sandqvist, T. P. Ray (Eds.); Central Activity in Galaxies. From Observational Data to Astrophysical Diagnostics. XIII, 235 pages. 1993.

Vol. 414: M. Napolitano, F. Sabetta (Eds.), Thirteenth International Conference on Numerical Methods in Fluid Dynamics. Proceedings, 1992. XIV, 541 pages. 1993.

Vol. 415: L. Garrido (Ed.), Complex Fluids. Proceedings, 1992. XIII, 413 pages. 1993.

Vol. 416: B. Baschek, G. Klare, J. Lequeux (Eds.), New Aspects of Magellanic Cloud Research. Proceedings, 1992. XIII, 494 pages. 1993.

Vol. 417: K. Goeke P. Kroll, H.-R. Petry (Eds.), Quark Cluster Dynamics. Proceedings, 1992. XI, 297 pages. 1993.

Vol. 418: J. van Paradijs, H. M. Maitzen (Eds.), Galactic High-Energy Astrophysics. XIII, 293 pages. 1993.

Vol. 419: K. H. Ploog, L. Tapfer (Eds.), Physics and Technology of Semiconductor Quantum Devices. Proceedings, 1992. VIII, 212 pages. 1993.

Vol. 420: F. Ehlotzky (Ed.), Fundamentals of Quantum Optics III. Proceedings, 1993. XII, 346 pages. 1993.

Vol. 421: H.-J. Röser, K. Meisenheimer (Eds.), Jets in Extragalactic Radio Sources. XX, 301 pages. 1993.

Vol. 422: L. Päivärinta, E. Somersalo (Eds.), Inverse Problems in Mathematical Physics. Proceedings, 1992. XVIII, 256 pages. 1993.

Vol. 423: F. J. Chinea, L. M. González-Romero (Eds.), Rotating Objects and Relativistic Physics. Proceedings, 1992. XII, 304 pages. 1993.

Vol. 424: G. F. Helminck (Ed.), Geometric and Quantum Aspects of Integrable Systems. Proceedings, 1992. IX, 224 pages. 1993.

Vol. 425: M. Dienes, M. Month, B. Strasser, S. Turner (Eds.), Frontiers of Particle Beams: Factories with $e^+ e^-$ Rings. Proceedings, 1992. IX, 414 pages. 1994.

Vol. 426: L. Mathelitsch, W. Plessas (Eds.), Substructures of Matter as Revealed with Electroweak Probes. Proceedings, 1993. XIV, 441 pages. 1994

Vol. 427: H. V. von Geramb (Ed.), Quantum Inversion Theory and Applications. Proceedings, 1993. VIII, 481 pages. 1994.

Vol. 429: J. L. Sanz, E. Martínez-González, L. Cayón (Eds.), Present and Future of the Cosmic Microwave Background. Proceedings, 1993. VIII, 233 pages. 1994.

New Series m: Monographs

Vol. m 1: H. Hora, Plasmas at High Temperature and Density. VIII, 442 pages. 1991.

Vol. m 2: P. Busch, P. J. Lahti, P. Mittelstaedt, The Quantum Theory of Measurement. XIII, 165 pages. 1991.

Vol. m 3: A. Heck, J. M. Perdang (Eds.), Applying Fractals in Astronomy. IX, 210 pages. 1991.

Vol. m 4: R. K. Zeytounian, Mécanique des fluides fondamentale. XV, 615 pages, 1991.

Vol. m 5: R. K. Zeytounian, Meteorological Fluid Dynamics. XI, 346 pages. 1991.

Vol. m 6: N. M. J. Woodhouse, Special Relativity. VIII, 86 pages. 1992.

Vol. m 7: G. Morandi, The Role of Topology in Classical and Quantum Physics. XIII, 239 pages. 1992.

Vol. m 8: D. Funaro, Polynomial Approximation of Differential Equations. X, 305 pages. 1992.

Vol. m 9: M. Namiki, Stochastic Quantization. X, 217 pages. 1992.

Vol. m 10: J. Hoppe, Lectures on Integrable Systems. VII, 111 pages. 1992.

Vol. m 11: A. D. Yaghjian, Relativistic Dynamics of a Charged Sphere. XII, 115 pages. 1992.

Vol. m 12: G. Esposito, Quantum Gravity, Quantum Cosmology and Lorentzian Geometries. Second Corrected and Enlarged Edition. XVIII, 349 pages. 1994.

Vol. m 13: M. Klein, A. Knauf, Classical Planar Scattering by Coulombic Potentials. V, 142 pages. 1992.

Vol. m 14: A. Lerda, Anyons. XI, 138 pages. 1992.

Vol. m 15: N. Peters, B. Rogg (Eds.), Reduced Kinetic Mechanisms for Applications in Combustion Systems. X, 360 pages. 1993.

Vol. m 16: P. Christe, M. Henkel, Introduction to Conformal Invariance and Its Applications to Critical Phenomena. XV, 260 pages. 1993.

Vol. m 17: M. Schoen, Computer Simulation of Condensed Phases in Complex Geometries. X, 136 pages. 1993.

Vol. m 18: H. Carmichael, An Open Systems Approach to Quantum Optics. X, 179 pages. 1993.

Vol. m 19: S. D. Bogan, M. K. Hinders, Interface Effects in Elastic Wave Scattering. XII, 182 pages. 1994.